# ASE Test Preparation Series

# Automobile Test

# Heating and Air Conditioning (Test A7)

## 4th Edition

DELMAR
CENGAGE Learning

Australia • Brazil • Japan • Korea • Mexico • Singapore • Spain • United Kingdom • United States

**ASE Test Preparation Series: Automobile Test: Heating and Air Conditioning (Test A7), Fourth Edition**

Vice President, Technology Professional Business Unit: Gregory L. Clayton

Product Development Manager: Kristen Davis

Product Manager: Kim Blakey

Editorial Assistant: Vanessa Carlson

Director of Marketing: Beth A. Lutz

Marketing Specialist: Brian McGrath

Marketing Coordinator: Marissa Maiella

Production Manager: Andrew Crouth

Production Editor: Kara A. DiCaterino

Senior Project Editor: Christopher Chien

XML Architect: Jean Kaplansky

Cover Design: Michael Egan

Cover Images: Portion courtesy of DaimlerChrysler Corporation

For product information and technology assistance, contact us at
**Cengage Learning Customer & Sales Support, 1-800-354-9706**
For permission to use material from this text or product,
submit all requests online at **www.cengage.com/permissions**
Further permissions questions can be emailed to
**permissionrequest@cengage.com**

ISBN-13: 978-1-4180-3884-7

ISBN-10: 1-4180-3884-9

**Delmar**
Executive Woods
5 Maxwell Drive
Clifton Park, NY 12065
USA

Cengage Learning is a leading provider of customized learning solutions with office locations around the globe, including Singapore, the United Kingdom, Australia, Mexico, Brazil, and Japan. Locate your local office at **www.cengage.com/global**

Cengage Learning products are represented in Canada by Nelson Education, Ltd.

To learn more about Delmar, visit **www.cengage.com/delmar**

Purchase any of our products at your local bookstore or at our preferred online store **www.ichapters.com**

**Notice to the Reader**

Printed in the United States of America
4 5 6 7 8      17 16 15 14 13

# Contents

# Section   5   Sample Test for Practice

# Section   6   Additional Test Questions for Practice

# Section   7   Appendices

# Preface

Delmar, Cengage Learning is very pleased that you have chosen our ASE Test Preparation Series to prepare yourself for the automotive ASE Examination. These guides are available for all of the automotive areas including A1–A8, the L1 Advanced Diagnostic Certification, the P2 Parts Specialist, the C1 Service Consultant and the X1 Undercar Specialist. These guides are designed to introduce you to the Task List for the test you are preparing to take, give you an understanding of what you are expected to be able to do in each task, and take you through sample test questions formatted in the same way the ASE tests are structured.

If you have a basic working knowledge of the discipline you are testing for, you will find Delmar's ASE Test Preparation Series to be an excellent way to understand the "must know" items to pass the test. These books are not textbooks. Their objective is to prepare the technician who has the requisite experience and schooling to challenge ASE testing. It cannot replace the hands-on experience or the theoretical knowledge required by ASE to master vehicle repair technology. If you are unable to understand more than a few of the questions and their explanations in this book, it could be that you require either more shop-floor experience or further study. Some resources that can assist you with further study are listed on the rear cover of this book.

Each book begins with an item-by-item overview of the ASE Task List with explanations of the minimum knowledge you must possess to answer questions related to the task. Following that there are 2 sets of sample questions followed by an answer key to each test and an explanation of the answers to each question. A few of the questions are not strictly ASE format but were included because they help teach a critical concept that will appear on the test. We suggest that you read the complete Task List Overview before taking the first sample test. After taking the first test, score yourself and read the explanation to any questions that you were not sure about, including the questions you answered correctly. Each test question has a reference back to the related task or tasks that it covers. This will help you to go back and read over any area of the task list that you are having trouble with. Once you are satisfied that you have all of your questions answered from the first sample test, take the additional tests and check them. If you pass these tests, you will be prepared to do well on the ASE test.

## Our Commitment to Excellence

The 4th edition of Delmar's ASE Test Preparation Series has been through a major revision with extensive updates to the ASE's task lists, test questions, and answers and explanations. Delmar, Cengage Learning has sought out the best technicians in the country to help with the updating and revision of each of the books in the series.

## About the Series Advisor

To promote consistency throughout the series, a series advisor took on the task of reading, editing, and helping each of our experts give each book the highest level of accuracy possible. Dan Perrin has served in the role of Series Advisor for the 4th edition of the ASE Test Preparation Series. Dan began ASE testing with the first series of tests in 1972 and has been continually certified ever since. He holds ASE master status in automotive, truck, collision, and machinist. He is also L1, L2, and alternated fuels certified, along with some others that have expired. He has been an automotive educator since 1979, having taught at the secondary, post-secondary, and industry levels. His service includes participation on boards that include the North American Council of Automotive Teachers (NACAT), the Automotive Industry Planning Council (AIPC), and the National Automotive Technicians Education Foundation (NATEF). Dan currently serves as the Executive Manager of NACAT and Director of the NACAT Education Foundation.

*Thanks for choosing Delmar's ASE Test Preparation Series. All of the writers, editors, Delmar Staff, and myself have worked very hard to make this series second to none. I know you are going to find this book accurate and easy to work with. It is our objective to constantly improve our product at Delmar by responding to feedback.*

*If you have any questions concerning the books in this series, you can email me at: autoexpert@trainingbay.com.*

*Dan Perrin*
*Series Advisor*

# The History and Purpose of ASE

ASE began as the National Institute for Automotive Service Excellence (NIASE). It was founded as a non-profit independent entity in 1972 by a group of industry leaders with the single goal of providing a means for consumers to distinguish between incompetent and competent technicians. It accomplishes this goal by testing and certification of repair and service professionals. From this beginning it has evolved to be known simply as ASE (Automotive Service Excellence) and today offers more than 40 certification exams in automotive, medium/heavy duty truck, collision, engine machinist, school bus, parts specialist, automobile service consultant, and other industry-related areas. At this time there are more than 400,000 professionals with current ASE certifications. These professionals are employed by new car and truck dealerships, independent garages, fleets, service stations, franchised service facilities, and more. ASE continues its mission by also providing information that helps consumers identify repair facilities that employ certified professionals through its Blue Seal of Excellence Recognition Program. Shops that have a minimum of 75% of their repair technicians ASE certified and meet other criteria can apply for and receive the Blue Seal of Excellence Recognition from ASE.

ASE recognized that educational programs serving the service and repair industry also needed a way to be recognized as having the faculty, facilities, and equipment to provide a quality education to students wanting to become service professionals. Through the combined efforts of ASE, industry, and education leaders, the non-profit National Automotive Technicians Education Foundation (NATEF) was created to evaluate and recognize training programs. Today more than 2000 programs are ASE certified under the standards set by the service industry. ASE/NATEF also has a certification of industry (factory) training program known as CASE. CASE stands for Continuing Automotive Service Education and recognizes training provided by replacement parts manufacturers as well as vehicle manufacturers.

ASE certification testing is administered by the American College Testing (ACT). Strict standards of security and supervision at the test centers insure that the technician who holds the certification earned it. Additionally ASE certification also requires that the person passing the test to be able to demonstrate that they have two years of work experience in the field before they can be certified. Test questions are developed by industry experts that are actually working in the field being tested. There is more detail on how the test is developed and administered in the next section. Paper and pencil tests are administered twice a year at over seven hundred locations in the United States. Computer based testing is now also available with the benefit of instant test results at certain established test centers. The certification is valid for five years and can be recertified by retesting. So that consumers can recognize certified technicians, ASE issues a jacket patch, certificate, and wallet card to certified technicians and makes signs available to facilities that employ ASE certified technicians.

You can contact ASE at any of the following:

National Institute for Automotive Service Excellence
101 Blue Seal Drive S.E.
Suite 101
Leesburg, VA 20175
Telephone 703-669-6600
FAX 703-669-6123
www.ase.com

WE SUPPORT
PROFESSIONAL CERTIFICATION
THROUGH THE
National Institute for
AUTOMOTIVE
SERVICE
EXCELLENCE

# 2 | Take and Pass Every ASE Test

Participating in an Automotive Service Excellence (ASE) voluntary certification program gives you a chance to show your customers that you have the "know-how" needed to work on today's modern vehicles. The ASE certification tests allow you to compare your skills and knowledge to the automotive service industry's standards for each specialty area.

If you are the "average" automotive technician taking this test, you are in your mid-thirties and have not attended school for about fifteen years. That means you probably have not taken a test in many years. Some of you, on the other hand, have attended college or taken postsecondary education courses and may be more familiar with taking tests and with test-taking strategies. There is, however, a difference in the ASE test you are preparing to take and the educational tests you may be accustomed to.

## How are the tests administered?

ASE test are administered at over 750 test sites in local communities. Paper and pencil tests are the type most widely available to technicians. Each tester is given a booklet containing questions with charts and diagrams where required. You can mark in this test booklet but no information entered in the booklet is scored. Answers are recorded on a separate answer sheet. You will enter your answers, using a number 2 pencil only. ASE recommends you bring four sharpened number 2 pencils that have erasers. Answer choices are recorded by coloring in the blocks on the answer sheet. The answer sheets are scanned electronically and the answers tabulated. For test security, test booklets include randomly generated questions. Your answer key must be matched to the proper booklet so it is important to correctly enter the booklet serial number on the answer sheet. All instructions are printed on the test materials and should be followed carefully.

ASE has introduced Computer Based Testing (CBT) at some locations. While the test content is the same for both testing methods the CBT tests have some unique requirements and advantages. It is strongly recommended that technicians considering the CBT tests go the ASE web page at www.ASE.com and review the conditions and requirements for this type of test. There is a demonstration of a CBT that allows you to experience this type of test before you register. Some technicians find this style of testing provides an advantage, while others find operating the computer a distraction. One significant benefit of CBT is the availability of instant results. You can receive your test results before you leave the test center. CBT testing also offers increased flexibility in scheduling. The cost for taking CBTs is slightly higher than paper and pencil tests and the number of testing sites is limited. The first time test taker may be more comfortable with the paper and pencil tests but technicians now have a choice.

## Who Writes the Questions?

The questions are written by service industry experts in the area being tested. Each area will have its own technical experts. Questions are entirely job related. They are designed to test the skills you need to be a successful technician. Theoretical knowledge is important and necessary to answer the questions, but the ability to apply that knowledge is the basis of ASE test questions.

Each question has its roots in an ASE "item-writing" workshop where service representatives from automobile manufacturers (domestic and import), aftermarket parts and equipment manufacturers, working technicians, and vocational educators meet in a workshop setting to share ideas and translate them into test questions. Each test question written by these experts must survive review by all members of the group.

The questions are written to deal with practical application of soft skills and system knowledge experienced by technicians in their day-to-day work.

All questions are pre-tested and quality-checked on a national sample of technicians. Those questions that meet ASE standards of quality and accuracy are included in the scored sections of the tests; the "rejects" are sent back to the drawing board or discarded altogether.

Each certification test is made up of between forty and eighty multiple-choice questions.

Note: Each test could contain additional questions that are included for statistical research purposes only. Your answers to these questions will not affect your score, but since you do not know which ones they are, you should answer all questions on the test. The five-year Recertification Test will cover the same content areas as those listed above. However, the number of questions in each content area of the Recertification Test will be reduced by about one-half.

## Objective Tests

A test is called an objective test if the same standards and conditions apply to everyone taking the test and there is only one correct answer to each question.

Objective tests primarily measure your ability to recall information. A well-designed objective test can also test your ability to understand, analyze, interpret, and apply your knowledge. Objective tests include true-false, multiple choice, fill in the blank, and matching questions. ASE's tests consist exclusively of four-part multiple-choice objective questions.

The following are some strategies that may be applied to your tests.

Before beginning to take an objective test, quickly look over the test to determine the number of questions, but do not try to read through all of the questions. In an ASE test, there are usually between forty and eighty questions, depending on the subject. Read through each question before marking your answer. Answer the questions in the order they appear on the test. Leave the questions blank that you are not sure of and move on to the next question. You can return to those unanswered questions after you have finished the others. They may be easier to answer at a later time after your mind has had additional time to consider them on a subconscious level. In addition, you might find information in other questions that will help you recall the answers to some of them.

Do not be obsessed by the apparent pattern of responses. For example, do not be influenced by a pattern like **D, C, B, A, D, C, B, A** on an ASE test.

There is also a lot of folk wisdom about taking objective tests. For example, there are those who would advise you to avoid response options that use certain words such as *all, none, always, never, must,* and *only,* to name a few. This, they claim, is because nothing in life is exclusive. They would advise you to choose response options that use words that allow for some exception, such as *sometimes, frequently, rarely, often, usually, seldom,* and *normally.* They would also advise you to avoid the first and last option (A and D) because test writers, they feel, are more comfortable if they put the correct answer in the middle (B and C) of the choices. Another recommendation often offered is to select the option that is either shorter or longer than the other three choices because it is more likely to be correct. Some would advise you to never change an answer since your first intuition is usually correct.

Although there may be a grain of truth in this folk wisdom, ASE test writers try to avoid them and so should you. There are just as many **A** answers as there are **B** answers, just as many **D** answers as **C** answers. As a matter of fact, ASE tries to balance the answers at about 25 percent per choice **A, B, C,** and **D**. There is no intention to use "tricky" words, such as outlined above. Put no credence in the opposing words "sometimes" and "never," for example.

Multiple-choice tests are sometimes challenging because there are often several choices that may seem possible, and it may be difficult to decide on the correct choice. The best strategy, in this case, is to first determine the correct answer before looking at the options. If you see the answer you decided on, you should still examine the options to make sure that none seem more correct than yours. If you do not know or are not sure of the answer, read each option very carefully and try to eliminate those options that you know to be wrong. That way, you can often arrive at the correct choice through a process of elimination.

If you have gone through all of the test and you still do not know the answer to some of the questions, <u>then guess.</u> Yes, guess. You then have at least a 25 percent chance of being correct. If you leave the question blank, you have no chance. Your score is based on the number of questions answered correctly.

# Preparing for the Exam

The main reason we have included so many sample and practice questions in this guide is, simply, to help you learn what you know and what you don't know. We recommend that you work your way through each question in this book. Before doing this, carefully look through Section 3; it contains a description and explanation of the question types you'll find on an ASE exam.

Once you understand what the questions will look like, move to the sample test. Answer one of the sample questions (Section 5) then read the explanation (Section 7) to the answer for that question. If you don't feel you understand the reasoning for the correct answer, go back and read the overview (Section 4) for the task that is related to that question. If you still don't feel you have a solid understanding of the material, identify a good source of information on the topic, such as a textbook, and do some more studying.

After you have completed all of the sample test items and reviewed your answers, move to the additional questions (Section 6). This time answer the questions as if you were taking an actual test. Do not use any reference or allow any interruptions in order to get a feel for how you will do on an actual test. Once you have answered all of the questions, grade your results using the answer key in Section 7. For every question that you gave a wrong answer to, study the explanations to the answers and/or the overview of the related task areas. Try to determine the root cause for your missing the question. The easiest thing to correct is learning the correct technical content. The hardest thing to correct is behaviors that lead you to a wrong conclusion. If you knew the information but still got it wrong there is a behavior problem that will need to be corrected. An example would be reading too quickly and skipping over words that affect your reasoning. If you can identify what you did that caused you to answer the question incorrectly you can eliminate that cause and improve your score. Here are some basic guidelines to follow while preparing for the exam:

- Focus your studies on those areas you are weak in.

- Be honest with yourself while determining if you understand something.

- Study often but in short periods of time.

- Remove yourself from all distractions while studying.

- Keep in mind the goal of studying is not just to pass the exam, the real goal is to learn!

- Prepare physically by getting a good night's rest before the test and eat meals that provide energy but do not cause discomfort.

- Arrive early to the test site to avoid long waits as test candidates check in and to allow all of the time available for your tests.

# During the Test

On paper and pencil tests you will be placing your answers on a sheet where you will be required to color in your answer choice. Stray marks or incomplete erasures may be picked up as an answer by the electronic reader, so be sure only your answers end up on the sheet. One of the biggest problems an adult faces in test taking, it seems, is placing the answer in the correct spot on the answer sheet. Make certain that you mark your answer for, say, question 21, in the space on the answer sheet designated for the answer for question 21. A correct response in the wrong line will probably result in two questions being marked wrong, one with two answers (which could include a correct answer but will be scored wrong) and the other with no answer. Remember, the answer sheet on the written test is machine scored and can only "read" what you have colored in.

If you finish answering all of the questions on a test and have remaining time, go back and review the answers to those questions that you were not sure of. You can often catch careless errors by using the remaining time to review your answers. Carefully check your answer sheet for blank answer blocks or missing information.

At practically every test, some technicians will invariably finish ahead of time and turn their papers in long before the final call. Some technicians may be doing recertification tests and others may be taking fewer tests than you. Do not let them distract or intimidate you.

It is not wise to use less than the total amount of time that you are allotted for a test. If there are any doubts, take the time for review. Any product can usually be made better with some additional effort. A test is no exception. It is not necessary to turn in your test paper until you are told to do so.

## Testing Time Length

An ASE written test session is four hours. You may attempt from one to a maximum of four tests in one session. It is recommended, however, that no more than a total of 225 questions be attempted at any test session. This will allow for just over one minute for each question.

Visitors are not permitted at any time. If you wish to leave the test room, for any reason, you must first ask permission. If you finish your test early and wish to leave, you are permitted to do so only during specified dismissal periods.

You should monitor your progress and set an arbitrary limit to how much time you will need for each question. This should be based on the number of questions you are attempting. It is suggested that you wear a watch because some facilities may not have a clock visible to all areas of the room.

Computer-Based Tests are allotted a testing time according to the number of questions ranging from one half hour to one and one half hours. Advanced level tests are allowed two hours. This time is by appointment and you should be sure to be on time to insure that you have all of the time allocated. If you arrive late for a CBT test appointment you will only have the amount of time remaining on your appointment.

## Your Test Results!

You can gain a better perspective about tests if you know and understand how they are scored. ASE's tests are scored by American College Testing (ACT), a nonpartial, unbiased organization having no vested interest in ASE or in the automotive industry.

Each question carries the same weight as any other question. For example, if there are fifty questions, each is worth 2 percent of the total score. The passing grade is 70 percent. That means you must correctly answer thirty-five of the fifty questions to pass the test.

The test results can tell you:

• where your knowledge equals or exceeds that needed for competent performance, or

• where you might need more preparation.

Your ASE test score report is divided into content areas and will show the number of questions in each content area and how many of your answers were correct. These numbers provide information about your performance in each area of the test. However, because there may be a different number of questions in each content area of the test, a high percentage of correct answers in an area with few questions may not offset a low percentage in an area with many questions.

It should be noted that one does not "fail" an ASE test. The technician who does not pass is simply told "More Preparation Needed." Though large differences in percentages may indicate problem areas, it is important to consider how many questions were asked in each area. Since each test evaluates all phases of the work involved in a service specialty, you should be prepared in each area. A low score in one area could keep you from passing an entire test.

There is no such thing as average. You cannot determine your overall test score by adding the percentages given for each task area and dividing by the number of areas. It doesn't work that way because there generally are not the same number of questions in each task area. A task area with twenty questions, for example, counts more toward your total score than a task area with ten questions.

Your test report should give you a good picture of your results and a better understanding of your strengths and weaknesses for each task area.

If you fail to pass the test, you may take it again at any time it is scheduled to be administered. You are the only one who will receive your test score. Test scores will not be given over the telephone by ASE nor will they be released to anyone without your written permission.

# 3 | Types of Questions on an ASE Exam

ASE certification tests are often thought of as being tricky. They may seem to be tricky if you do not completely understand what is being asked. The following examples will help you recognize certain types of ASE questions and avoid common errors.

Paper-and-pencil tests and computer-based test questions are identical in content and difficulty. Most initial certification tests are made up of forty to eighty multiple-choice questions. Multiple-choice questions are an efficient way to test knowledge. To answer them correctly, you must think about each choice as a possibility, and then choose the one that best answers the question. To do this, read each word of the question carefully. Do not assume you know what the question is about until you have finished reading it.

About 10 percent of the questions on an actual ASE exam will use an illustration. These drawings contain the information needed to correctly answer the question. The illustration must be studied carefully before attempting to answer the question. Often, technicians look at the possible answers then try to match up the answers with the drawing. Always do the opposite; match the drawing to the answers. When the illustration is showing an electrical schematic or another system in detail, look over the system and try to figure out how the system works before you look at the question and the possible answers.

## Multiple-Choice Questions

The most common type of question used on ASE Tests is the multiple-choice question. This type of question contains three "distracters" (wrong answers) and one "key" (correct answer). When the questions are written effort is made to make the distracters plausible to draw an inexperienced technician to one of them. This type of question gives a clear indication of the technician's knowledge. Using multiple criteria including cross-sections by age, race, and other background information, ASE is able to guarantee that a question does not bias for or against any particular group. A question that shows bias toward any particular group is discarded. If you encounter a question that you are unsure of, reverse engineer it by eliminating the items that it cannot be. For example:

A rocker panel is a structural member of which vehicle construction type?

A. Front-wheel drive
B. Pickup truck
C. Unibody
D. Full-frame

Analysis:

This question asks for a specific answer. By carefully reading the question, you will find that it asks for a construction type that uses the rocker panel as a structural part of the vehicle.

Answer A is wrong. Front-wheel drive is not a vehicle construction type.
Answer B is wrong. A pickup truck is not a type of vehicle construction.
**Answer C is correct.** Unibody design creates structural integrity by

welding parts together, such as the rocker panels, but does not require exterior cosmetic panels installed for full strength.
Therefore, the correct answer is C. If the question was read quickly and the words "construction type" were passed over, answer A may have been selected.
Answer D is wrong. Full-frame describes a body-over-frame construction type that relies on the frame assembly for structural integrity.
Therefore, the correct answer is C. If the question was read quickly and the words "construction type" were passed over, answer A may have been selected.

# EXCEPT Questions

Another type of question used on ASE tests has answers that are all correct except one. The correct answer for this type of question is the answer that is wrong. The word "**EXCEPT**" will always be in capital letters. You must identify which of the choices is the wrong answer. If you read quickly through the question, you may overlook what the question is asking and answer the question with the first correct statement. This will make your answer wrong. An example of this type of question and the analysis is as follows:

All of the following are tools for the analysis of structural damage **EXCEPT**:

A. height gauge
B. tape measure.
C. dial indicator.
D. tram gauge.

Analysis:

The question really requires you to identify the tool that is not used for analyzing structural damage. All tools given in the choices are used for analyzing structural damage except one. This question presents two basic problems for the test-taker who reads through the question too quickly. It may be possible to read over the word "**EXCEPT**" in the question or not think about which type of damage analysis would use answer C. In either case, the correct answer may not be selected. To correctly answer this question, you should know what tools are used for the analysis of structural damage. If you cannot immediately recognize the incorrect tool, you should be able to identify it by analyzing the other choices.

Answer A is wrong. A height gauge may be used to analyze structural damage.
Answer B is wrong. A tape measure may be used to analyze structural damage.
**Answer C is correct.** A dial indicator may be used as a damage analysis tool for moving parts, such as wheels, wheel hubs, and axle shafts, but would not be used to measure structural damage.
Answer D is wrong. A tram gauge is used to measure structural damage.

# Technician A, Technician B Questions

The type of question that is most popularly associated with an ASE test is the "Technician A says . . . Technician B says . . . Who is right?" type. In this type of question, you must identify the correct statement or statements. To answer this type of question correctly, you must carefully read each technician's statement and judge it on its own merit to determine if the statement is true.

Sometimes this type of question begins with a statement about some analysis or repair procedure. This is often referred to as the stem of the question and provides the setup or background information required to understand the conditions the question is based on. This is followed by two statements about the cause of the concern, proper inspection, identification, or repair choices. You are asked whether the first statement, the second statement, both statements, or neither statement is correct.

Analyzing this type of question is a little easier than the other types because there are only two ideas to consider although there are still four choices for an answer.

Technician A, Technician B questions are really double true or false questions. The best way to analyze this kind of question is to consider each technician's statement separately. Ask yourself, is A true or false? Is B true or false? Then select your answer from the four choices. An important point to remember is that an ASE Technician A, Technician B question will never have Technician A and B directly disagreeing with each other. That is why you must evaluate each statement independently.

An example of this type of question and the analysis of it follows.

A vehicle comes into the shop with a gas gauge that will not register above one half full. When the sending unit circuit is disconnected the gauge reads empty and when it is connected to ground the gauge goes to full. Technician A says that the sending unit is shorted to ground. Technician B says the gauge circuit is working and the sending unit is likely the problem. Who is right?

A. A only
B. B only
C. Both A and B
D. Neither A nor B

Analysis:

Reading of the stem of the question sets the conditions of the customer concern and establishes what information is gained from testing. General knowledge of gauge circuits and test procedures are needed to correctly evaluate the technician's conclusions. Note: Avoid being distracted by experience with unusual or problem vehicles that you may have worked on, Other technicians taking the same test do not have that knowledge, so it should not be used as the basis of your answers.

Technician A is wrong because a shorted to ground sending unit would produce a gauge reading equivalent to the test conditions of a grounding the circuit and produce a full reading.
**Technician B is correct** because the gauge spans when going from an open circuit to a completely
grounded circuit. This would tend to indicate that the problem had to be in the sending unit.
Answer C is not correct. Both technicians are identifying the problem as a sending unit but technician A qualified the problem as a specific type of failure (grounded) that would not have caused the symptoms of the vehicle.
Answer D is not correct because technician B's diagnosis is a possible cause of the conditions identified.

# Most-Likely Questions

Most-Likely questions are somewhat difficult because only one choice is correct while the other three choices are nearly correct. An example of a Most-Likely-cause question is as follows:

The Most-Likely cause of reduced turbocharger boost pressure may be a:

A. wastegate valve stuck closed.
B. wastegate valve stuck open.
C. leaking wastegate diaphragm.
D. disconnected wastegate linkage.

Analysis:

Answer A is wrong. A wastegate valve stuck closed increases turbocharger boost pressure.

**Answer B is correct.** A wastegate valve stuck open decreases turbocharger boost pressure.
Answer C is wrong. A leaking wastegate valve diaphragm increases turbocharger boost pressure.
Answer D is wrong. A disconnected wastegate valve linkage will increase turbocharger boost pressure.

## LEAST-Likely Questions

Notice that in Most-Likely questions there is no capitalization. This is not so with LEAST-Likely type questions. For this type of question, look for the choice that would be the LEAST-Likely cause of the described situation. Read the entire question carefully before choosing your answer. An example is as follows:

What is the LEAST-Likely cause of a bent pushrod?

A. Excessive engine speed
B. A sticking valve
C. Excessive valve guide clearance
D. A worn rocker arm stud

Analysis:

Answer A is wrong. Excessive engine speed may cause a bent pushrod.
Answer B is wrong. A sticking valve may cause a bent pushrod.
**Answer C is correct.** Excessive valve clearance will not generally cause a bent pushrod.
Answer D is wrong. A worn rocker arm stud may cause a bent pushrod.

You should avoid relating questions to those unusual situations that you may have encountered and answer based on the technical and mechanical possibilities.

## Summary

There are no four-part multiple-choice ASE questions having "none of the above" or "all of the above" choices. ASE does not use other types of questions, such as fill-in-the-blank, completion, true-false, word-matching, or essay. ASE does not require you to draw diagrams or sketches. If a formula or chart is required to answer a question, it is provided for you. There are no ASE questions that require you to use a pocket calculator.

# Overview of the Task List

## Heating and Air Conditioning (Test A7)

The following section includes the task areas and task lists for this test and a written overview of the topics covered in the test.

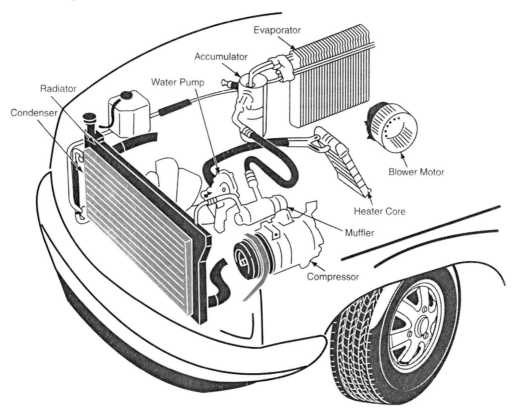

### An Automotive Air Conditioning Primer

Since many of you may be approaching air conditioning for the first time, you may not have the same basic understanding of what goes on inside the system as you do for engines or brakes. To avoid having to assemble your knowledge from the task list like a jigsaw puzzle, we are providing a basic explanation of air-conditioning theory.

Refer to the provided diagram to make yourself familiar with the systems operation. The basic principal of any air-conditioning system is to take advantage of the temperature drop that occurs during evaporation of a liquid. When you get out of the shower, still wet, and feel breezes you did not feel when you were dry, you are experiencing evaporation. In the automotive air-conditioning environment, we use the same principal along with a heat exchanger so that we can reuse the same liquid over and over.

Let's start at the heart of the system you know as the compressor. The compressor has a suction and a discharge side. The compressor does not handle liquid refrigerant. It pumps it as a gas. Compressors may use small pistons and connecting rods that hook to the main shaft of the compressor, or they may be rotary/vane type that work very similar to rotary engines. In both examples, each cylinder or chamber alternates between creating vacuum and pressure that is controlled by reed valves. There are some variations, but this is the most common design.

The liquid in the air-conditioning system is called refrigerant. All refrigerants have a very low boiling point, which is to say that they change from a liquid to a gas below ambient temperature unless they are under pressure. They also evaporate very quickly which is why they work so well to make things cold.

When the warm gaseous refrigerant is pumped from the compressor, it goes to the condenser where a heat exchange process occurs that turns the gas back into a liquid. The liquid refrigerant may pass through a filter with a desiccant in it that will attract water and particles that do not belong in the system. These driers, or accumulators, are holding tanks for the liquid refrigerant. Ahead of this part of the system, which is often referred to as the "high" or discharge side, is a control unit of one type or another. This unit may be in the form of an orifice tube or expansion valve. Both serve the same purpose: to stop straight liquid refrigerant from entering and flooding the evaporator. As you can tell by its name, this is where the cold air is created. The restriction in the system allows the liquid refrigerant to be sprayed into the evaporator where it quickly evaporates and returns to its gaseous form again. The air that the blower circulates across the evaporator is cooled by evaporation and heat exchange, which is why you see water under a vehicle when the air conditioner has been working hard. The water was in the incoming air, but condenses on the evaporator and is carried out of the evaporator housing by a drain. The compressor draws up the refrigerant and the cycle starts again.

There are controls on the system that regulate pressure, cabin temperature, and even the compressor itself. We will cover these in each task area as they come up. The task list describes the actual work you should be able to do as a technician that you will be tested on by the ASE. This is your key to the test and you should review this section carefully. We have based our sample test and additional questions upon these tasks. The overview section will also support your understanding of the task list. ASE advises that the questions on the test may not equal the number of tasks listed; the task lists tell you what ASE expects you to know how to do and be ready to be tested upon.

At the end of each question in the Sample Test and Additional Test Questions sections, a letter and number will be used as a reference back to this section for additional study. Note the following example: **A.4.**

# A.  A/C System Service, Diagnosis, and Repair (12 Questions)

**Task A.4**    **Leak test A/C system; determine needed repairs.**

Example:
1. An oily residue is present on the fittings of the hose connected from the compressor to the condenser. Technician A says this residue may be caused by excessive oil in the refrigerant system. Technician B says this residue may be caused by a leak at the hose fittings. Who is right?
    A. Technician A only
    B. Technician B only
    C. Both A and B
    D. Neither A nor B.                                                (A.4)

Analysis:

Question #1
Answer A is wrong. An excessive oil charge would cause elevated system pressures and warmer duct temperatures but it would not cause oil to leak out of hose and line connections.
**Answer B is correct.** Only Technician B is correct. An oily residue indicates a refrigerant leak in that area where refrigerant is escaping along with some of the oil in the system.
Answer C is wrong. Only Technician B is correct.
Answer D is wrong. Only Technician B is correct.

# Task List and Overview

## A.  A/C System Service, Diagnosis, and Repair (12 Questions)

**Task A.1**   **Identify system type and conduct performance test on the A/C system; determine needed repairs.**

Performance testing provides a measure of air-conditioning system operating efficiency. A manifold pressure gauge set is used to determine both high and low pressures in the refrigeration system. The desired pressure readings will vary according to temperature. Use temperature/pressure charts as a guide to determine the proper pressures. At the same time, a thermometer is used to determine air discharge temperature into the passenger compartment. Before making this test, it should be established that the blend door, which controls airflow over the heater core and the evaporator, is moving completely to the cool position and/or that the heater control valve (if equipped) is closing completely.

The typical procedure for a performance test follows:

1. Connect the manifold gauge set to the respective high- and low-pressure fittings. These fittings are found in various locations within the high- and low-pressure sides of the system.

2. Close the hood and all of the doors and windows of the vehicle.

3. Adjust the air-conditioning controls to maximum cooling and high blower position.

4. Idle the engine in neutral or park with the brake on. For the best results, place a high-volume fan in front of the radiator grille to insure an adequate supply of airflow across the condenser.

5. Increase engine speed to 1,500 to 2,000 rpm.

6. Measure the temperature at the evaporator air outlet grille or air duct nozzle (35° to 40° F).

7. Read the high and low pressures, and compare them to the normal range of the operating pressure given in the service manual.

Operating pressures vary with humidity as well as with outside air temperature. Accordingly, on more humid days, operating pressures will be on the high side of the range indicated in the service manual's performance chart. On less humid days, the operating pressures will read toward the lower side. If operating pressures are found to be within the normal range, the refrigeration portion of the air-conditioning system is functioning properly. This can be further confirmed with a check of evaporator outlet air temperatures.

Always refer to the pressure charts given in the service manual when basing your diagnosis of the system on system pressures. Although the specifications will vary, here are some guidelines to help you interpret abnormal readings.

• If the high-side pressure is too high, suspect air in the system, too much refrigerant in the system, a restriction in the high side of the system, and poor airflow across the condenser.

• If the high-side pressure is too low, suspect a low refrigerant level or defective compressor.

• If the low-side pressure is higher than normal, suspect refrigerant overcharge, a defective compressor, or faulty metering device.

• If the low-side pressure is lower than normal, suspect a faulty metering device, poor airflow across the evaporator, a restriction in the low side of the system, or the system is undercharged with refrigerant.

Evaporator outlet air temperature also varies according to outside (ambient) air and humidity conditions. Further variations can be found, depending on whether the system is controlled by a cycling clutch compressor or an evaporator pressure control valve. Because of these variations, it is difficult to pinpoint what evaporator outlet air temperature should be on all applications. In general, with low-side air temperatures (70°F) and humidity (20 percent), the evaporator outlet air temperature should be in the 35- to 40-degree range. On the other extreme of 80°F outside air temperatures and 90 percent humidity condition, the evaporator air outlet temperature might be in the 55- to 60-degree range.

Since it is impractical to provide a specific performance chart for all the different types of air-conditioning systems, it is desirable to develop an experience factor for determining the correlation that can be anticipated between operating pressures and outlet air temperatures on the various systems. For example, feel the discharge line from the compressor to the condenser. The discharge line should be the same temperature along its full length. Any temperature change is a sign of restriction, and the line should be flushed or replaced. Perform this test carefully because the discharge line will be hot.

There are other tests that can be performed with the engine running.

- Check the condenser by feeling up and down the face or along the return bends for a temperature change. There should be a gradual change from hot to warm as you go from the top to the bottom. Any abrupt change indicates a restriction, and the condenser has to be flushed or replaced.

- If the system has a receiver/drier, check it. The inlet and outlet lines should be the same temperature. Any temperature difference or frost on the lines or receiver tank are signs of a restriction. The receiver/drier must be replaced.

- If the system has a sight glass, check it as previously described.

- Feel the liquid line from the receiver/drier to the expansion valve. The line should be warm for its entire length.

- The expansion valve should be free of frost, and there should be a sharp temperature difference between its inlet and outlet.

- The suction line to the compressor should be cool to the touch from the evaporator to the compressor. If it is covered with thick frost, this might indicate that the expansion valve is flooding the evaporator.

- Typically, the formation of frost on the outside of a line or component means there is a restriction to the flow of refrigerant.

- On vehicles equipped with the orifice tube system, feel the liquid line from the condenser outlet to the evaporator inlet. A restriction is indicated by any temperature change in the liquid line before the crimp dimples the orifice tube in the evaporator inlet. Flush the liquid line or replace the orifice tube if restricted.

- The accumulator as well as the suction line must be cool to the touch from the evaporator outlet to the compressor.

By combining the results of both the hands-on checks and an interpretation of pressure gauge readings, the technician has a good indication that some unit in the system is malfunctioning and that further diagnosis is needed.

## Task A.2    Diagnose A/C system problems indicated by pressure gauge and/or temperature readings; determine needed repairs.

A restriction in the low side will cause lower-than-normal low-side pressure, even into a vacuum. Such a restriction in the low side of the system would generally be accompanied by a lower-than-normal high-side pressure. This is also true if there is an undercharge of refrigerant. A restriction in the high side of the system will result in a higher-than-normal high-side pressure. This may generally be accompanied by a lower-than-normal low-side pressure. If the system is overcharged, however, the higher-than-normal high-side pressure will be accompanied by a higher-than-normal low-side pressure.

Air and moisture in the refrigerant system or a restricted thermal expansion valve (TXV) cause higher-than-specified system pressures. A defective compressor may cause low high-side pressure and high low-side pressure. A low refrigerant charge causes reduced low-side and high-side pressures.

A restricted TXV may cause low refrigerant pressures. This condition may also result in TXV frosting. Air in the refrigerant system may cause high system pressures. The manifold gauge set is one of the most valuable air conditioning tools. It is used when discharging, charging, evacuating, and for diagnosing the system. With the new legislation on handling refrigerants, all gauge sets are required to have a valve device to close off the end of the hose so that the fitting not in use is automatically shut.

The low-pressure gauge is graduated into pounds of pressure from 1 to 120 (with cushion to 250) in 1-pound graduations, and, in the opposite direction, in inches of vacuum from 0 to 30. This is the gauge that should always be used in checking pressure on the low-pressure side of the system. The gauge at the right is graduated from 0 to 500 pounds pressure in 10-pound graduations. This is the high-pressure gauge that is used for checking pressure on the high-pressure side of the system.

The center manifold fitting is common to both the low and the high side and is for evacuating or adding refrigerant to the system. When this fitting is not being used, it should be capped. A test hose connected to the fitting directly under the low-side gauge is used to connect the low side of the test manifold to the low side of the system. A similar connection is found on the high side.

The gauge manifold is designed to control refrigerant flow. When the manifold test set is connected into the system, pressure is registered on both gauges at all times. During all tests, both the low- and high-side hand valves are in the closed position (turned inward until the valve is seated).

Refrigerant flows around the valve stem to the respective gauges and registers the system's low-side pressure on the low-side gauge and the system's high-side pressure on the high-side gauge. The hand valves isolate the low and high side from the central portion of the manifold. When the gauges are first connected to the gauge fittings with the refrigeration system charged, the gauge lines should always be purged. Purging is done by cracking each valve on the gauge set to allow the pressure of the refrigerant in the refrigeration system to force the air to escape through the center gauge line. Failure to purge lines can result in air or other contaminants entering the refrigeration system.

NEVER open the high-side hand valve with the system operating and a refrigerant can connected to the center hose. The refrigerant will flow out of the system under high pressure into the can. High-side pressure is between 150 and 300 psi and will cause the refrigerant tank to burst. The only occasion for opening both hand valves at the same time would be when evacuating the system or when reclaiming refrigerant with the system off.

Because R-134A is not interchangeable with R-12, separate sets of hoses, gauges, and other equipment are required to service vehicles. All equipment used to service R-134A and R-12 systems must meet SAE standard J1991. The service hoses on the manifold gauge set must have manual or automatic backflow valves at the service port connector ends. This prevents the refrigerant from being released into the atmosphere during connection and disconnection. Manifold gauge sets for R-134A can be identified by one or all of the following: Labeled *FOR USE WITH R-134A*, Labeled *HFC-134* or *R-134A*, and/or have a light blue color on the face of the gauges.

For identification purposes, R-134A service hoses must have a black stripe along their length and be clearly labeled *SAE J2196/R-134A*. The low-pressure hose is blue with a black stripe. The high-pressure hose is red with a black stripe. The center service hose is yellow with a black stripe. Service hoses for one type of refrigerant will not easily connect into the wrong system, as the fittings for an R-134A system are different than those used in an R-12 system.

## Task A.3    Diagnose A/C system problems indicated by sight, sound, smell, and touch procedures; determine needed repairs.

The thermostatic expansion valve is the coldest part in the system, but heavy ice buildup is not normal. A restriction in the receiver/drier would produce ice at that spot. The outlet of the metering device is the coldest part in the system. It is normal for frost or a light accumulation of ice to occur at the outlet of the metering device. If there is a restriction in the system, there would be heavy frost or ice buildup at the point of the restriction. If, for example, the receiver/drier were clogged, there would be ice buildup at the outlet of the receiver/drier.

A restricted receiver/drier usually causes frosting of this component. A flooded evaporator causes frosting of the evaporator outlet and compressor suction pipes. A restricted TXV usually causes frosting of this component. A low refrigerant charge may cause faster than normal clutch cycling without frosting of any components.

## Task A.4    Leak test A/C system; determine needed repairs.

Testing the refrigerant system for leaks is one of the most important phases of troubleshooting. Over a period of time, all air conditioners lose or leak some refrigerant. In systems that are in good condition, refrigerant losses of up to one-half pound per year are considered normal. Higher loss rates signal a need to locate and repair the leaks.

Leaks are most often found at the compressor hose connections and at the various fittings and joints in the system. Refrigerant can be lost through hose permeation. Leaks can also be traced to pinholes in the evaporator caused by acid, which forms when water and refrigerant mix. Since oil and refrigerant leak out together, oily spots on hoses, fittings, and components mean there is a leak. Any suspected leak should be confirmed by using any one of these methods of detection. All leaks should be repaired before refrigerant is added to the system.

As mentioned before, the presence of oil around the fitting of an air-conditioning line or hose is an indication of a refrigerant leak. Carefully check all system connections. This method of leak detection is the easiest to conduct but also the least effective.

Electronic leak detection is one method of leak detection; it is safe, effective, and can be used with all types of refrigerants. The hand-held battery-operated electronic leak detector contains a test probe that is moved about one inch per second in areas of suspected leaks. (Remember that refrigerant gas is heavier than air, thus the probe should be positioned below the test point.) An alarm or a buzzer on the detector indicates the presence of a leak. On some models, a light flashes to establish the leak.

To find a refrigerant leak using the fluorescent tracer system, first introduce a fluorescent dye into the air-conditioning system by means of a special infuser. Run the air conditioner for a few minutes, giving the tracer dye fluid time to circulate and penetrate. Wear the tracer protective goggles and scan the system with a black-light glow gun. Leaks in the system will shine under the black light as a luminous yellow-green.

Leaks can also be located by applying leak detector fluid around areas to be tested. There are dyes available for R-12 and R-134A systems; make sure you use the correct one for the refrigerant for which you are looking. If a leak is present, it will form clusters of bubbles around the source. A very small leak will cause white foam to form around the leak source within several seconds to a minute. Adequate lighting over the entire surface being tested is necessary for an accurate diagnosis.

Regardless of which type of leak detector is used, the system should have a minimum of 70-psi pressure to accurately detect a leak. Be sure to check the entire system. If a leak is found at a connection, tighten the connection carefully and recheck. If the leak is still apparent, discharge the system. Replace the damaged components, evacuate, charge, and test the system for leaks.

## Task A.5    Identify A/C system refrigerant.

On older automotive air-conditioning systems, the refrigerant was Refrigerant-12 (commonly referred to as R-12 and Freon). R-12 is dichlorodifluoromethane ($CCl_2F_2$). An amendment to the United States' Clean Air Act and international agreements has mandated that R-12 be phased out as the refrigerant of choice. This came as a result of research that found the earth's ozone layer was being deteriorated by the chemicals found in R-12. The ozone layer is the earth's outermost shield of protection. This delicate layer protects against harmful effects of the sun's ultraviolet rays. The thinning of the ozone layer has become a worldwide concern. The ozone depletion is caused in part by release of chlorofluorocarbons (CFCs) into the atmosphere. R-12 is in the chemical family of CFCs. Since air-conditioning systems with R-12 are susceptible to leaks, further damage to the ozone layer could be avoided by not using R-12 in air-conditioning units.

Of the many chemicals that could be used in place of R-12, the automobile manufacturers have decided to use R-134A. This refrigerant may also be referred to as SUVA. R-134A is a hydrofluorocarbon (HFC) that causes less damage to the ozone layer when released to the atmosphere.

Although R-134A air conditioners operate in the same way and with the same basic components as R-12 systems, the two refrigerants are not interchangeable. Because it is less efficient than R-12, R-134A must operate at higher pressures to make up for the loss of performance and requires new service techniques and system component designs. Basically, the higher system pressures of R-134A mean the system must be designed for those higher pressures.

Always use the same type of refrigerant and refrigerant oil for which the system was designed. If the system is being converted to accept an alternative refrigerant, make sure all old refrigerant and oil is removed from the system. Also make sure the vehicle is clearly labeled as to the type of refrigerant it was converted to use. The EPA has approved a number of refrigerants, other than R-134A. These can be used, but the system must be modified to contain and effectively use these refrigerants. There are several disadvantages to using these alternative refrigerants. One is based on the future. If the system has been converted to use an alternative refrigerant and the vehicle needs service in the future, will the alternative refrigerant be available at that shop or at all? Another disadvantage is many air conditioning part manufacturers will not warrant their parts if something other than R-12 or R-134A is used.

Keep in mind that all EPA-accepted refrigerants that substitute for R-12, except R-134A, are blends that contain ozone-depleting HCFCs. Therefore, the venting of any of these refrigerants into the atmosphere is prohibited.

Before servicing a system unknown to you, it is a very wise idea to use a refrigerant identification machine that will tell you the content of the system before potentially contaminating your machine with blended refrigerants. With the significant amounts of misinformation that was and still is present in some circles, there are many "retrofitted" systems out there that have everything from R-12 oils and R-134A refrigerant, to some having pure butane in them. Protect your equipment and yourself.

It must never be assumed that a system is charged with R-12 simply because it has 1/4-inch service valves. It should never be assumed that the lack of a label or labels is an indication that the system is charged with R-12 refrigerant. R-12 is sold in white containers, and R-134A is marketed in light blue containers.

## Task A.6     Evacuate A/C system.

All of the refrigerant in the system must be recovered prior to evacuation. Evacuation is the name given to the process that pulls all traces of air, moisture, and refrigerant from the system. This is done by creating a vacuum in the system. A vacuum pump is connected to the system to do this. The vacuum pump should remain on and connected to the system for at least 30 minutes after 26 to 29 inches of mercury is reached.

Any air or moisture that is left inside an air-conditioning system reduces the system's efficiency and eventually leads to major problems, such as compressor failure.

Air causes excessive pressure within the system, restricting the refrigerant's ability to change its state from gas to liquid within the refrigeration cycle, which drastically reduces its heat absorbing and transferring ability. Moisture, on the other hand, can cause freeze-up at the cap tube or expansion valve, which restricts refrigerant flow or blocks it completely. Both of these problems result in intermittent cooling or no cooling at all. Moisture also forms hydrochloric acid when mixed with refrigerant, causing internal corrosion, that is especially dangerous to the compressor.

The vacuum pump reduces system pressure in order to vaporize the moisture and then exhausts the vapor along with all remaining air. The pump's ability to clean the system is directly related to its ability to reduce pressure–create a vacuum–low enough to boil off all the contaminating moisture.

An electronic thermistor vacuum gauge is designed to work with the vacuum pump to measure the last, most critical inch of mercury during evacuation. It constantly monitors and visually indicates the vacuum level so you know when the system has a full vacuum and will be moisture-free. After the system is evacuated, it can be recharged. If the system will not pull down to a good vacuum, there is probably a leak somewhere in the system.

## Task A.7     Inspect system components for contamination.

Various components of the AC system can become contaminated. Some common sources of contamination include metal and other material from a failing compressor, desiccant particles from a ruptured desiccant container in the receiver/drier or accumulator, and sludge from moisture and oil emulsification. These contaminants often collect at points of restriction such as the orifice tube or expansion valve. They also collect at filter locations such as the outlet of the receiver/drier and the oil by-pass passage filter in the accumulator. Contamination will usually result in a system performance loss. A failing compressor loses effectiveness but can also cause contamination blockage at the mentioned locations. If gauge readings and system performance checks indicate a compressor problem, the system should be inspected for contamination. Even after compressor replacement and a thorough flushing of the system, some contamination may still be in the system. Many replacement compressor manufacturers recommend the installation of an in-line filter after compressor replacement. Desiccant and sludge contamination usually results in a restriction in the system indicated by an unusual temperature change at the point of restriction. A complete understanding of the relationship between system components and the normal system pressures and temperatures will allow the technician to easily locate the restriction by sense of touch. Correction requires disassembly to determine the source of the restriction. Sludge contamination indicates the system should be flushed of all oil and evacuated completely and have the receiver/drier or accumulator changed. Desiccant contamination dictates the replacement of drier or accumulator.

**Task A.8    Charge A/C system with refrigerant (liquid or vapor).**

The importance of the correct charge cannot be stressed enough. The efficient operation of the air-conditioning system greatly depends on the correct amount of refrigerant in the system. A low charge results in inadequate cooling under high heat loads, due to a lack of reserve refrigerant, and can cause the clutch cycling switch to cycle faster than normal. An overcharge can cause inadequate cooling because of a high liquid refrigerant level in the condenser. Refrigerant controls will not operate properly and compressor damage can result. In general, an overcharge of refrigerant will cause higher-than-normal gauge readings and noisy compressor operation.

The charging cylinder is designed to meter out a desired amount of a specific refrigerant by weight. Compensation for temperature variations is accomplished by reading the pressure on the gauge of the cylinder and dialing the plastic shroud. The calibrated chart on the shroud contains corresponding pressure readings for the refrigerant being used.

When charging an air-conditioning system with refrigerant, the pressure in the system often reaches a point at which it is equal to the pressure in the cylinder from which the system is being charged. To get more refrigerant into the system to complete the charge, heat must be applied to the cylinder.

The A/C system may only be charged through the high side with the system off. Inverting the refrigerant container disperses liquid refrigerant. As a general practice, the system is vapor charged through the low side while it is running. If the system is charged through the high side (with the system off), the compressor should be turned a few times by hand afterward to ensure there is no liquid refrigerant on top of the piston. If liquid refrigerant enters the compressor, this component may be damaged. The vehicle should always be charged from the low-pressure side if the engine is on. If the engine is off, it can be charged from the high-pressure side.

**Task A.9    Identify AC system lubricant type and capacity.**

Normally the only source of lubrication for a compressor is the oil mixed with the refrigerant. Because of the loads and speeds at which the compressor operates, proper lubrication is a must for long compressor life. The refrigerant oil required by the system depends on a number of things, but it is primarily dictated by the refrigerant used in the system. R-12 systems use a mineral oil. Mineral oil mixes well with R-12 without breaking down. Mineral oil, however, cannot be used with R-134A. R-134A systems require a synthetic oil: polyalkylene glycol (PAG). There are a number of different blends of PAG oil; always use the one recommended by the vehicle manufacturer or compressor manufacturer. Failure to use the correct oil may cause damage to the compressor.

Most often when a system has been converted from R-12 to R-134A, ester oil will be recommended. Since there may be some mineral oil and R-12 still in the system, it is important to use oil that will mix well with the residue and not break down. Ester oil mixes well with mineral oil because it is hydrocarbon-based.

When replacing refrigerant oil, it is important to use the specific type and quantity of oil recommended by the compressor manufacturer. If there is a surplus of oil in the system, too much oil circulates with the refrigerant, causing the cooling capacity of the system to be reduced. Too little oil results in poor lubrication of the compressor. When there has been excessive leakage or it is necessary to replace a component of the refrigeration system, certain procedures must be followed to assure that the total oil charge in the system is correct after leak repair or the new part is on the car.

When the compressor is operated, oil gradually leaves the compressor and is circulated through the system with the refrigerant. Eventually a balanced condition is reached in which a certain amount of oil is retained in the compressor and a certain amount is continually circulated. If a component of the system is replaced after the system has been operated, some oil goes with it. To maintain the original total oil charge, it is necessary to compensate for this by adding oil to the new replacement part. Because of the differences in compressor designs, be sure to follow the manufacturer's instructions when adding refrigerant oil to their unit. The only way to know the exact amount of oil in the system is to remove all of the oil from the system and recharge with the correct amount specified by the equipment manufacturer. This can be accomplished by draining the compressor and flushing the system with an approved flush agent. It will be necessary to remove the system restriction and flush the system components individually.

### Task A.10      Inspect and replace passenger compartment (cabin air, pollen) filter.

A cabin air filter is used on the inlet of the air distribution system to clean foreign materials from the air before it reaches the evaporator. Two types of filters are common: filter paper and absorption. Both can remove very small particles from the inlet air including pollen, mold spores, and other allergens that can affect passenger health and comfort. These filters are usually located under the hood at the outside air intake point, under the dash, or in both places. Replacement should be at scheduled maintenance intervals. Driving conditions, geographic location, and system usage will alter the frequency of needed service. Typically, the recommended replacement interval is between 15,000 and 24,000 miles. A visual inspection is the standard way to evaluate the need for replacement. If the filter becomes clogged, reduced airflow through the evaporator can affect system performance and possibly result in evaporator freeze-up.

## B.  Refrigeration System Component Diagnosis and Repair (10 Questions)

### Task B.1      Compressor and Clutch (5 Questions)

### Task B.1.1      Diagnose A/C system problems that cause the protection devices (pressure, thermal, and electronic controls) to interrupt system operation; determine needed repairs.

The ambient temperature switch senses outside air temperature and is designed to prevent compressor clutch engagement when air conditioning is not required or when compressor operation might cause internal damage to seals and other parts. The switch is in series with the compressor clutch electrical circuit and closes at about 37° F. At all lower temperatures, the switch is open, preventing clutch engagement.

In cycling clutch systems, a thermostatic switch is placed in series with the compressor clutch circuit so it can turn the clutch on or off. It de-energizes the clutch and stops the compressor if the evaporator is at the freezing point. When the temperature of the evaporator approaches the freezing point, the thermostatic switch opens the circuit and disengages the compressor clutch. The compressor remains inoperative until the evaporator temperature rises to the preset temperature, at which time the switch closes and compressor operation resumes.

A pressure cycling switch is electrically connected in series with the compressor electromagnetic clutch. Like the thermostatic switch, the turning on and off of the pressure cycling switch controls the operation of the compressor.

The low-pressure cutoff or discharge pressure switch is located on the high side of the system and senses any low-pressure conditions. It is tied into the compressor clutch circuit, allowing it to immediately disengage the clutch when the pressure falls too low.

The electronic cycling clutch switch prevents evaporator freeze-up by sending a signal to the engine control computer. The computer, in turn, cycles the compressor on and off by monitoring suction line temperature. If the temperature gets too low, the ECCS will open the input circuit to the computer, which causes the A/C clutch relay to open. In turn, this disengages the compressor clutch. Often this switch is the thermostatic switch at the evaporator.

Many manufacturers are using air conditioning pressure sensors in place of high and low pressure switches. These pressure transducers are typically located in the high side of the system and provide a varying voltage to the ECM/PCM as the pressure changes. These pressure sensors have three terminals and operate much like a MAP sensor with the ECM/PCM sending a reference voltage and providing a sensor ground on two of the terminals. The third terminal is the signal and changes its voltage level as the system pressure changes. This input to the ECM/PCM can be seen in the data stream of most scan tools while monitoring engine data.

The high-pressure relief valve is used to keep system pressures from reaching a point that may cause compressor lock-up or other component damage because of excessive high pressures. When system pressures exceed a predetermined point, the pressure relief valve opens, reducing the system's pressure.

**Task B.1.2    Inspect, test, and replace A/C system pressure and thermal protection devices.**

If there are 12 volts at the compressor clutch, there is battery voltage supplied to the compressor clutch and connecting wire. If the clutch is inoperative, the compressor clutch coil, the wire from the coil, or the wire from the coil to the fuse is open. An open thermal switch may also be the cause.

In some A/C systems, a thermal switch is connected in series with the compressor clutch. This switch usually is mounted on the compressor. Many thermal switches open at 257°F (125°C) and close at 230°F (110°C).

**Task B.1.3    Inspect, adjust, and replace A/C compressor drive belts, pulleys, and tensioners.**

Always use the exact replacement size of belt. The size of a new belt is typically given, along with the part number, on the belt container. After replacing a belt, make sure it is adjusted properly. Some engines have an adjusting bolt that can be tightened to bring the belt tension to specifications. On other engines, it may be necessary to use a pry bar to move an accessory enough to meet tension specs. Be careful not to damage the part you are prying against. The belt's tension should be checked with a belt tension gauge.

When installing a serpentine belt, make sure it is fed in and around the accessories properly. Service manuals and under-hood decals show the proper belt routing. Also make sure the belt tensioner is working properly. After any drive belt has been installed with the correct tension, tighten any bolts or nuts that were loosened to move the belt.

A loose A/C compressor belt will likely produce a squealing noise when the compressor is clutch-engaged. Since the air pump does not require much power to turn it, this belt is not likely to produce a squealing noise on acceleration. Because the power steering pump is always under load, a loose power steering belt may cause a squealing noise on acceleration. Pulleys can become worn and cause belt misalignment. Frequently pulleys that run against the back of serpentine belts are made of plastic and will show wear.

V-belt pulleys will also wear and result in belt noise on correctly tensioned new belts. A visual inspection will usually identify worn, bent, or misaligned pulleys. Replacement should follow manufacturers' recommended procedures and specifications.

**Task B.1.4    Inspect, test, service, and replace A/C compressor clutch components or assembly.**

Every compressor is equipped with an electromagnetic clutch as part of the compressor pulley assembly. It is designed to engage the pulley to the compressor shaft when the clutch coil is energized. The purpose of the clutch is to transmit power from the engine to the compressor and to provide a means of engaging and disengaging the refrigeration system from engine operation.

The clutch is driven by power from the engine's crankshaft, which is transmitted through one or more belts (a few use gears) to the pulley, which is in operation whenever the engine is running. When the clutch is engaged, power is transmitted from the pulley to the compressor shaft by the clutch drive plate. When the clutch is not engaged, the compressor shaft does not rotate, and the pulley freewheels.

The clutch is engaged by a magnetic field and disengaged by springs when the magnetic field is broken. When the controls call for compressor operation, the electrical circuit to the clutch is completed, the magnetic clutch is energized, and the clutch engages the compressor. When the electrical circuit is opened, the clutch disengages the compressor.

Two types of electromagnetic clutches have been in use for many years. Early-model air-conditioning systems used a rotating coil clutch. The magnetic coil, which engages or disengages the compressor, is mounted within the pulley and rotates with it. Electrical connections for the clutch operation are made through a stationary brush assembly and rotating slip rings that are part of the field coil assembly. This older rotating coil clutch, now in limited use, has been largely replaced by the stationary coil clutch. The compressor clutch can be checked with an ohmmeter. Service manuals will give the acceptable resistance for the field coil of the clutch assembly.

With the stationary coil, wear has been measurably reduced, efficiency increased, and serviceability made much easier. The clutch coil does not rotate. When the driver first energizes the air-conditioning system from the passenger compartment dashboard, the pulley assembly is magnetized by the stationary coil on the compressor body, thus engaging the clutch to the clutch hub that is attached to the compressor shaft. This activates the air conditioning system. Depending on the system, the magnetic clutch is usually pressure controlled to cycle the operation of the compressor (depending on system temperature or pressure). In some system designs, the clutch might operate continually when the system is turned on. With stationary coil design, service is not usually necessary except for an occasional check on the electrical connections.

Nearly all clutch assemblies have a clearance spec of .020 to .030 of an inch for the distance between the clutch and the pressure plate. This clearance is measured with a feeler gauge. If the clearance is too great, the clutch may slip and cause a scraping or squealing noise. If the clearance is insufficient, the compressor may run when not electrically activated and the clutch may chatter at all times.

## Task B.1.5    Identify required lubricant type; inspect and correct level in A/C compressor.

Normally the only source of lubrication for a compressor is the oil mixed with the refrigerant. Because of the loads and speeds at which the compressor operates, proper lubrication is a must for long compressor life. The refrigerant oil required by the system depends on a number of things, but it is primarily dictated by the refrigerant used in the system. R-12 systems use a mineral oil. Mineral oil mixes well with R-12 without breaking down. Mineral oil, however, cannot be used with R-134A. R-134A systems require a synthetic oil, polyalkylene glycol (PAG). There are a number of different blends of PAG oil; always use the one recommended by the vehicle manufacturer or compressor manufacturer. Failure to use the correct oil will cause damage to the compressor.

Generally, compressor oil level is checked only when there is evidence of a major loss of system oil that could be caused by a broken refrigerant hose, severe hose fitting leak, badly leaking compressor seal, or collision damage to the system's components.

When replacing refrigerant oil, it is important to use the specific type and quantity of oil recommended by the compressor manufacturer. If there is a surplus of oil in the system, too much oil circulates with the refrigerant, causing the cooling capacity of the system to be reduced. Too little oil results in poor lubrication of the compressor. When there has been excessive leakage or it is necessary to replace a component of the refrigeration system, certain procedures must be followed to assure that the total oil charge in the system is correct after leak repair or the new part is on the car.

When the compressor is operated, oil gradually leaves the compressor and is circulated through the system with the refrigerant. Eventually a balanced condition is reached in which a certain amount of oil is retained in the compressor and a certain amount is continually circulated. If a component of the system is replaced after the system has been operated, some oil goes with it. To maintain the original total oil charge, it is necessary to compensate for this by adding oil to the new replacement part. Because of the differences in compressor designs, be sure to follow the manufacturer's instructions when adding refrigerant oil to their unit.

When replacing an A/C compressor, the recommended procedure is to measure the oil recovered from the system while discharging. Drain and measure the oil left in the old compressor. Drain the new compressor and refill it with the same amount of oil removed from the system and old compressor. If 2 oz. (59 ml) of oil are recovered from the system and 2 oz. (59 ml) of oil drained from the old compressor, drain the replacement compressor and then add 4 oz. (118 ml) of oil to the compressor before installation.

A small amount of lubricant is lost from the compressor due to circulation through the system. The lubricant must be drained from some compressors and measured in a beaker. The lubricant level in some compressors may be measured with a dipstick while others must be removed, drained, and measured in a beaker. The lubricant used in automotive air-conditioning systems is a non-foaming, sulfur-free grade specially formulated for use in certain types of air-conditioning systems. It must be noted that a mineral-based lubricant is used in R-12 systems while a glycol-based synthetic lubricant is used in an R-134A system.

## Task B.1.6    Inspect, test, service or replace A/C compressor.

The compressor is the heart of the automotive air-conditioning system. It separates the high-pressure and low-pressure sides of the system. The primary purpose of the unit is to draw the low-pressure and low-temperature vapor from the evaporator and compress this vapor into high-temperature, high-pressure vapor. This action results in the refrigerant having a higher temperature than surrounding air and enables the condenser to condense the vapor back to a liquid. The secondary purpose of the compressor is to circulate or pump the refrigerant through the A/C system under the different pressures required for proper operation.

A piston compressor can have its pistons arranged in an in-line, axial, radial, or V design. It is designed to have an intake stroke and a compression stroke for each cylinder. On the intake stroke, the refrigerant from the low side of the system (evaporator) is drawn into the compressor. The intake of refrigerant occurs through intake reed valves. These one-way valves control the flow of refrigerant vapors into the cylinder. During the compression stroke, the vaporous refrigerant is compressed. This increases both the pressure and the temperature of the heat-carrying refrigerant. The outlet or discharge-side reed valves then open to allow the refrigerant to move to the condenser. The outlet reed valves are the beginning of the high side of the system. Reed valves are made of spring steel that can be weakened or broken if improper charging procedures are used, such as liquid charging with the engine running.

A common variation of a piston-type compressor is the variable displacement compressor. These compressors not only act as a regular compressor but they also control the amount of refrigerant that passes through the evaporator. The pistons are connected to a wobble-plate. The angle of the wobble-plate determines the stroke of the pistons and is controlled by the difference in pressure between the outlet and inlet of the compressor. When the stroke of the pistons is increased, more refrigerant is being pumped and there is increased cooling.

The rotary vane compressor does not have pistons. It consists of a rotor with several vanes and a carefully shaped housing. As the compressor shaft rotates, the vanes and housing form chambers. The refrigerant is drawn through the suction port into these chambers, which become smaller as the rotor turns. The discharge port is located at the point where the gas is completely compressed. No sealing rings are used in a vane compressor. The vanes are sealed against the housing by centrifugal force and lubricating oil. The oil sump is located on the discharge side, so the high pressure tends to force it around the vanes into the low-pressure side. This action ensures continuous lubrication. Because this type of compressor depends on a good oil supply, it is subject to damage if the system charge is lost. A protection device is used to disengage the clutch if pressure drops too low.

The scroll-type compressor has a movable scroll and a fixed or non-movable scroll that provide an eccentric-like motion. As the compressor's crankshaft rotates, the movable scroll forces the refrigerant against the fixed scroll and toward the center of the compressor. This motion pressurizes the refrigerant. The action of a scroll-type compressor can be compared to that of a tornado. The pressure of air moving in a circular pattern increases as it moves toward the center of the circle. A delivery port is positioned at the center of the compressor and allows the high-pressure refrigerant to flow into the air-conditioning system. This type of compressor operates more smoothly than other designs.

When replacing a compressor, it is important to not alter the oil capacity of the A/C system in the process. Always follow the directions that come with each compressor about how to handle the oil for the new compressor. If there are no specific directions with the new compressor, then follow the steps listed here. Drain the oil from the old compressor into a measuring cup. Drain the oil from the new compressor and discard the lubricant. If more than two ounces of oil were drained from the old compressor, then add that amount of new lubricant to the new compressor. If less than two ounces of oil were drained from the old compressor, then add two ounces of new lubricant to the new compressor. Also, make sure to add the correct amount of lubricant to the other components that get replaced such as the drier or accumulator. It is also advisable to add lubricant to any devices that were flushed during the repair such as the condenser or evaporator. Flushing these components strips the lubricant from the surface and there is no longer lubricant throughout the system.

## Task B.1.7    Inspect, repair or replace A/C compressor mountings.

A compressor noise problem is often difficult to localize. When the compressor is operating, a loose compressor mount or bracket may vibrate, giving the impression that the compressor itself is somehow defective. This type of noise usually stops when the compressor is disengaged.

If the compressor mount is loose, this can cause the compressor to move and make noise. If the A/C system does not have any refrigerant in it, the compressor clutch will not engage and, therefore, will not make any noise.

## Task B.2   Evaporator, Condenser, and Related Components (5 Questions)

## Task B.2.1   Inspect, repair, or replace A/C system mufflers, hoses, lines, filters, fittings, and seals.

An air-conditioning system is divided into two sides: the high side and the low side. High side refers to the side of the system that is under high pressure and high temperature. Low side refers to the low-pressure, low-temperature side of the system.

All the major components of the system have inlet and outlet connections that accommodate either flare or O-ring fittings. The refrigerant lines that connect between these units are made up of an appropriate length of hose or tubing with flare or O-ring fittings at each end as required. In either case, the hose or tube end of the fitting is constructed with sealing beads to accommodate a hose or tube clamp connection.

There are three major refrigerant lines. Suction lines are located between the outlet side of the evaporator and the inlet side or suction side of the compressor. They carry the low-pressure, low-temperature refrigerant vapor to the compressor where it again is recycled through the system. Suction lines are always distinguished from the discharge lines by touch and size. They are cold to the touch. The suction line is larger in diameter than the liquid line because refrigerant in a vapor state takes up more room than refrigerant in a liquid state.

The other types of refrigerant lines are the discharge and liquid lines. Beginning at the discharge outlet on the compressor, the discharge or high-pressure line connects the compressor to the condenser. The liquid lines connect the condenser to the receiver/drier and the receiver/drier to the inlet side of the expansion valve. Through these lines, the refrigerant travels in its path from a gas state (compressor outlet) to a liquid state (condenser outlet) and then to the inlet side of the expansion valve where it vaporizes on entry to the evaporator. Discharge and liquid lines are always very warm to the touch and easily distinguishable from the suction lines.

Aluminum tubing is commonly used to connect air-conditioning components where flexibility is not required. Where the line is subjected to vibrations, special rubber hoses are used. Typically the compressor outlet and inlet lines are rubber hoses with aluminum ends and fittings.

R-134A systems are required to be fitted with quick-disconnect fittings through the system. These also have hoses specially made for R-134A. They have an additional layer of rubber that serves as a barrier to prevent the refrigerant from escaping through the pores of the hose. Some late-model R-12 systems also use these barrier hoses to prevent the loss of refrigerant through the walls of the hoses.

Some refrigerant line connections are sealed with O-rings and retained with spring lock couplings. A special tool is used to release the spring lock couplings.

It is not necessary to replace the O-ring seals unless they are leaking. The accumulation of dirt around an A/C line connection is an indication of a leak in the system. When there is a leak in the system, dirt collects in the area of the leak due to some of the oil leaking out.

## Task B.2.2   Inspect A/C condenser for airflow restrictions.

The condenser consists of coiled refrigerant tubing mounted in a series of thin cooling fins to provide maximum heat transfer in a minimum amount of space. The condenser is normally mounted just in front of the vehicle's radiator. It receives the full flow of ram air from the movement of the vehicle or airflow from the radiator fan when the vehicle is standing still.

The purpose of the condenser is to condense or liquefy the high-pressure, high-temperature vapor coming from the compressor. To do so, it must give up its heat. The condenser receives very hot (normally 200°F to 400°F), high-pressure refrigerant vapor from the compressor through its discharge hose. The refrigerant vapor enters the inlet at the top of the condenser, and as the hot vapor passes down through the condenser coils, heat (following its natural tendencies) moves from the hot refrigerant into the cooler air as it flows across the condenser coils and fins. This process causes a large quantity of heat to be transferred to the outside air and the refrigerant to change from a high-pressure hot vapor to a high-pressure warm liquid. This high-pressure warm liquid flows from the outlet at the bottom of the condenser through a line to the receiver/drier or to the refrigerant metering device if an accumulator instead of a drier is used.

In an air-conditioning system, which is operating under an average heat load, the condenser has a combination of hot refrigerant vapor in the upper two-thirds of its coils. The lower third of the coils contains the warm liquid refrigerant, which has condensed. This high-pressure liquid refrigerant flows from the condenser and on toward the evaporator. In effect, the condenser is a true heat exchanger.

Ram airflow across the condenser is produced by vehicle movement or the action of the fan. A defective fan clutch generally causes the fan to run at maximum potential at all times, which will usually increase ram airflow. Reduced airflow across the condenser generally affects the ability of the condenser to cool the refrigerant, resulting in higher temperatures. The higher refrigerant temperatures increase suction pressure in the system, affecting system cooling abilities.

If the condenser air passages are severely restricted, this may result in high refrigerant system pressures and refrigerant discharge from the high-pressure relief valve.

Restrictions will not allow enough air to cool down the condenser, causing higher temperatures and in turn higher pressures.

## Task B.2.3    Inspect, test, and replace A/C system condenser and mountings.

If the system were overcharged with refrigerant, the head pressure may be excessive but there would be no frosting at the condenser outlet. If the condenser is restricted, there would be frosting at the point of restriction. Frosting near the condenser outlet is a clear indication that there is a restriction in the condenser.

When frost is forming on one of the condenser tubes near the bottom of the condenser, the refrigerant passage is restricted at that location in the condenser. Replacement of the condenser requires recovery of the refrigerant, removal by disconnecting inlet and outlet connections, measurement and replacement of the correct type and quantity of oil, and recharging. The difficulty of the task varies from manufacturer to manufacturer.

## Task B.2.4    Inspect and replace receiver/drier or accumulator/drier.

Used on many early systems, the receiver/drier is a storage tank for the liquid refrigerant from the condenser, which flows into the upper portion of the receiver tank containing a bag of desiccant (moisture-absorbing material such as silica alumina or silica gel). As the refrigerant flows through an opening in the lower portion of the receiver, it is filtered through a mesh screen attached to a baffle at the bottom of the receiver. The desiccant in this assembly is to absorb any moisture present that might enter the system during assembly. These features of the assembly prevent obstruction to the valves or damage to the compressor.

Depending on the manufacturer, the receiver/drier may be known by other names such as filter or dehydrator. Regardless of its name, the function is the same. Included in many receiver/driers are additional features such as high-pressure fitting, a pressure relief valve, and a sight glass for determining the state and condition of the refrigerant in the system.

The receiver/drier is often neglected when the air-conditioning system is serviced or repaired. Failure to replace it can lead to poor system performance or replacement part failure. It is recommended that the receiver/drier and/or its desiccant be changed whenever a component is replaced, the system has lost the refrigerant charge, or the system has been open to the atmosphere for any length of time.

The desiccant draws moisture like a magnet and it can become contaminated in less than five minutes if it is exposed to the atmosphere. Keep it sealed. Once opened, put it under a vacuum quickly.

Many late-model systems are not equipped with a receiver/drier; rather, they use an accumulator to accomplish the same thing. The accumulator is connected into the low side, at the outlet of the evaporator. The accumulator also contains a desiccant and is designed to store excess refrigerant and to filter and dry the refrigerant. If liquid refrigerant flows out of the evaporator, it will be collected by and stored in the accumulator. The main purpose of an accumulator is to prevent liquid from entering the compressor.

## Task B.2.5    Inspect, test, and replace expansion valve(s).

The refrigerant flow to the evaporator must be controlled to obtain maximum cooling, while ensuring complete evaporation of the liquid refrigerant within the evaporator. This is accomplished by a thermostatic expansion valve (TEV or TXV) or a fixed orifice tube.

The TEV is mounted at the inlet to the evaporator and separates the high-pressure side of the system from the low-pressure side. The TEV regulates refrigerant flow to the evaporator to prevent evaporator flooding or starving. In operation, the TEV regulates the refrigerant flow to the evaporator by balancing the inlet flow to the outlet temperature.

Both externally and internally equalized TEVs are used in air-conditioning systems. The only difference between the two valves is that the external TEV uses an equalizer line connected to the evaporator outlet line as a means of sensing evaporator outlet pressure. The internal TEV senses evaporator inlet pressure through an internal equalizer passage. Both valves have a capillary tube to sense evaporator outlet temperature. The tube is filled with a gas, which, if allowed to escape due to careless handling, will ruin the TXV.

During stabilized conditions, the pressure on the bottom of the expansion valve diaphragm becomes equal to the pressure on the top of the diaphragm. This allows the valve spring to close the valve. When the system is started, the pressure on the bottom of the diaphragm drops rapidly, allowing the valve to open and meter liquid refrigerant to the lower evaporator tubes where it begins to vaporize.

Compressor suction draws the vaporized refrigerant out of the top of the evaporator at the top tube where it passes by the sealed sensing bulb. The bottom of the valve diaphragm internally senses the evaporator pressure through the internal equalization passage around the sealed sensing bulb. As evaporator pressure is increased, the diaphragm flexes upward pulling the pushrod away from the ball seat of the expansion valve. The expansion valve spring forces the ball onto the tapered seat and the liquid refrigerant flow is reduced.

As the pressure is reduced due to restricted refrigerant flow, the diaphragm flexes downward again, opening the expansion valve to provide the required controlled pressure and refrigerant flow condition. As the cool refrigerant passes by the body of the sensing bulb, the gas above the diaphragm contracts and allows the expansion valve spring to close the expansion valve. When heat from the passenger compartment is absorbed by the refrigerant, it causes the gas to expand. The pushrod again forces the expansion valve to open, allowing more refrigerant to flow so that more heat can be absorbed.

A defective capillary tube would most likely cause the TXV to remain closed, causing poor cooling at all times. An overcharged system will cause higher suction pressures and poor cooling at all times. A restricted condenser passage will cause low cooling at all times. Some symptoms indicate moisture freezing in the TXV. When the TXV freezes, refrigerant flow is reduced and cooling is also reduced. As the TXV thaws, refrigerant flow is restored and cooling is also restored until the TXV freezes again.

## Task B.2.6    Inspect and replace orifice tube(s).

Like the thermostatic expansion valve, the orifice tube is the dividing point between the high- and low-pressure parts of the system. However, its metering or flow rate control does not depend on comparing evaporator pressure and temperature. It is a fixed orifice. The flow rate is determined by pressure difference across the orifice and by subcooling. Subcooling is additional cooling of the refrigerant in the bottom of the condenser after it has changed from vapor to liquid. The flow rate through the orifice is more sensitive to subcooling than to pressure difference.

When using the special tool to remove an orifice tube, pour a small amount of refrigerant oil on top of the tube and engage the notch in the tool in the orifice tube. Hold the T-handle and rotate the outer sleeve to remove the orifice tube.

A restricted orifice tube may cause lower-than-specified low-side pressure. A restricted orifice tube may cause frosting of the tube because of the low temperatures caused by the low pressure.

## Task B.2.7    Inspect, test, clean, or replace evaporator(s).

The evaporator, like the condenser, consists of a refrigerant coil mounted in a series of thin cooling fins. It provides a maximum amount of heat transfer in a minimum amount of space. The evaporator is usually located beneath the dashboard or instrument panel.

Upon receiving the low-pressure, low-temperature liquid refrigerant from the thermostatic expansion valve or orifice tube in the form of an atomized (or droplet) spray, the evaporator serves as a boiler or vaporizer. This regulated flow of refrigerant boils immediately. Heat from the core surface is lost to the boiling and vaporizing refrigerant, which is cooler than the core, thereby cooling the core. The air passing over the evaporator loses its heat to the cooler surface of the core, thereby cooling the air inside the car. As the process of heat loss from air to the evaporator core surface is taking place, any moisture (humidity) in the air condenses on the outside of the evaporator core and is drained off as water. A drain tube in the bottom of the evaporator housing leads the water outside the vehicle. This dehumidification of air is an added feature of the air-conditioning system that adds to passenger comfort. It can also be used as a means of controlling fogging of the vehicle windows. Under certain conditions, however, too much moisture can accumulate on the evaporator coils. An example would be when humidity is extremely high and the maximum cooling mode is selected. The evaporator temperature might become so low that moisture would freeze on the evaporator coils before it can drain off.

Through the metering, or controlling, action of the thermostatic expansion valve or orifice tube, greater or lesser amounts of refrigerant are provided in the evaporator to adequately cool the car under all heat load conditions. If too much refrigerant is allowed to enter, the evaporator floods. This results in poor cooling due to the higher pressure (and temperature) of the refrigerant. The refrigerant can neither boil away rapidly nor vaporize. On the other hand, if too little refrigerant is metered, the evaporator starves.

Poor cooling again results because the refrigerant boils away or vaporizes too quickly before passing through the evaporator. The temperature of the refrigerant vapor at the evaporator outlet will be approximately 4°F to 16°F higher than the temperature of the liquid refrigerant at the evaporator inlet. This temperature differential is the superheat that ensures that the vapor will not contain any droplets of liquid refrigerant that would be harmful to the compressor.

## Task B.2.8    Inspect, clean, and repair evaporator housing and water drain.

A plugged evaporator case drain may cause windshield fogging, but this problem would not result in an oily film on the windshield. An oily film on the windshield may be caused by a refrigeration leak in the evaporator core that may allow some refrigerant oil to escape in the evaporator case.

Remove the evaporator core from the case for proper cleaning. Most evaporators have no filter; therefore, airborne debris, such as lint, hair, and other contaminants, enters the evaporator case and clings to the wet evaporator core. It is nearly impossible to adequately clean an evaporator core without first removing it from the case.

The evaporator water drain is an integral part of the evaporator housing assembly. A crack or break in the housing may reduce the cooling capacity of the air conditioner. If the evaporator (water) drain becomes clogged, it is easily cleaned with a stiff wire from the outlet in, provided the technician takes care not to puncture the delicate evaporator core. If the case has a crack or break on the engine side of the firewall, excessive engine heat may enter the case and thereby reduce the efficiency of the air-conditioning system.

## Task B.2.9    Inspect, test, and replace evaporator pressure/temperature control systems and devices.

Evaporator controls maintain a backpressure in the evaporator. Because of the refrigerant temperature/pressure relationship, the effect is to regulate evaporator temperature. The temperature is controlled to a point that provides effective air cooling but prevents the freezing of moisture that condenses on the evaporator.

In this type of system, the compressor operates continually when dash controls are in the air-conditioning position. Evaporator outlet air temperature is automatically controlled by an evaporator pressure control valve. These types of valves throttle the flow of refrigerant out of the evaporator as required to establish a minimum evaporator pressure and thereby prevent freezing of condensation on the evaporator core.

In some refrigerant systems with an evaporator pressure regulator (EPR), pilot-operated absolute valve (POA), or suction throttling valve (STV), the compressor runs continually in the A/C mode. Excessive pressure drop across the EPR valve indicates this valve is sticking closed.

For some time, the (ETR) evaporator temperature regulator and suction throttling valve (STV) were popular methods of evaporator temperature control, but these have not been in use for several years. All were used in the suction line from the evaporator to the compressor to maintain evaporator pressure, thereby maintaining its temperature.

### Task B.2.10 Identify, inspect, and replace A/C system service valves (gauge connections) and valve caps.

System service valves, which incorporate a service gauge port for manifold gauge set connections, are provided on the low and high sides of most air-conditioning systems. When making gauge connections, purge the gauge lines first by cracking the charging valve and allowing a small amount of refrigerant to flow through the lines. Then, connect the lines immediately.

Most systems have Schrader service valves in both the high and low portions of the system for test purposes. Closely resembling a tire valve, Schrader valves are usually located in the high-pressure line (from compressor to condenser) and in the low-pressure line (from evaporator to condenser) to permit checking of the high side and low side of the system. All test hoses have a Schrader core depressor in them. As the hose is threaded into the service port, the pin in the center of the valve is depressed, allowing refrigerant to flow to the manifold gauge set. When the hose is removed, the valve closes automatically.

After disconnecting gauge lines, check the valve areas to be sure the service valves are correctly seated and the Schrader valves are not leaking. This last step can save you from having to recharge a system for free. This is where an electronic leak detector can come in very handy. After servicing a system, just check the service ports for leaks before replacing the caps.

R134a systems use unique fittings that help prevent cross-contamination of refrigerants. The low-side fitting is a 13mm quick-coupling design and the high-side fitting is a 16mm quick-coupling design.

### Task B.2.11 Inspect and replace A/C system high pressure relief device.

The high-pressure relief valve is used to keep system pressures from reaching a point that may cause compressor lockup or other component damage because of excessive high pressures. When system pressures exceed a predetermined point, the pressure relief valve opens, allowing excessive system pressures to be reduced. A high-pressure relief valve discharges refrigerant at approximately 475 psi (3,275 kpa).

If replacing the high-pressure relief valve, the system must be discharged before removing the valve. It is necessary to replace the high-pressure relief valve with one of the same relief pressure specifications unless the vehicle is going to be retrofitted to use R-134A refrigerant instead of R-12.

Some high-pressure relief devices are spring-loaded and self-resetting. Some high-pressure relief devices are made of fusible metal and are not self-resetting. Self-resetting high-pressure relief valves are usually found on the rear head of the compressor while fusible metal high-pressure relief devices are often found on the receiver/drier, accumulator, or some other pressure vessel. Not all systems, however, have a pressure relief device; many aftermarket systems do not have one.

## C. Heating and Engine Cooling Systems Diagnosis and Repair (5 Questions)

### Task C.1 Diagnose the cause of temperature control problems in the heater/ventilation system; determine needed repairs.

In many cases, problems with the heating system are problems with the engine's cooling system. Therefore most service work and diagnosis is done to the cooling system. Problems that pertain specifically to the heater are few: the heater control valve and the heater. Most often, if these two items are faulty, the engine's cooling system will be negatively affected. Both of these items are replaced, rather than repaired. In some cases, it is possible to make repairs to vacuum hoses and electrical connections without removing the heater assembly. If it is necessary to remove the heater assembly, the cooling system must be drained before removing the heater core.

When diagnosing a heating system that is inadequate, it is imperative that you follow a logical diagnostic routine. The first objective is to be sure there is adequate coolant in the cooling system. Next, verify that the engine reaches normal operating temperature. In many cases, you will find a thermostat is staying open and your diagnosis is complete. If you have a way to see coolant flow from the upper hose into the radiator, you will see flow through the hose even when the engine is cool if the thermostat is open. If the engine is reaching operating temperature and the heater output is not acceptable, the next step is to check to see that there is flow through the heater core. Some systems use a heater control valve. These valves may be controlled by vacuum, electricity or a cable. Be sure that the valve changes position, allowing coolant into the heater core. If that area is good or not applicable, check the inlet and outlet temperatures or the heater hoses at the core. If one is cold, you have a restricted core. If both are hot and equal, you may have restricted fins on the outside of the core (usually from debris like leaves) or the blend door may not be opening to allow the blower to move air across the core. There will be some drop in outlet temperature if everything is working normally. Usually this would be 10–30 degrees. A drop more than that means the core is restricted. On the other side of the coin, if a system is always hot, even when on the cold setting, you must check the control valve if there is one and the blend door operation. ATC systems often use computer-controlled blend doors with position sensors that can fail.

Many ATC systems have a self-diagnostic feature that is pressing a series of buttons on the control head. This diagnostic feature will display trouble codes for failures within the system. Some systems will also recalibrate the doors while running self-diagnostics.

## Task C.2   Diagnose window fogging problems; determine needed repairs.

A sticky film on the inside of the windshield may be caused by a coolant leak in the heater core. Under this condition, the coolant level in the radiator should be checked. Repair or replace the heater core. Windshield fogging may be caused by a plugged air conditioning/heater case drain that allows water to collect in the case. This water can become stagnant and produce a very pungent odor in the passenger compartment. The windshield will not fog due to a low coolant level.

A leaking heater core is often the cause of a windshield fogging problem. Windshield fogging is the result of hot and humid air that condenses on the cooler glass. Cleaning a clogged evaporator/heater case drain tube often eliminates a windshield fogging problem. If the drain is clogged, a leaking heater core will allow engine coolant to accumulate in the housing. A clogged drain also allows condensed water vapor to accumulate. Both of these problems add moisture to the air, which tends to fog the windshield.

## Task C.3   Perform engine cooling system tests; determine needed repairs.

Cooling systems have to perform several functions and require tests for all of those functions. Probably the first and most important thing it must do is actually hold coolant. Leaks are the most common complaint with cooling systems. Since the system experiences considerable expansion and contraction, components will eventually become stressed and develop cracks or leaks. The first level of leak diagnosis is to visually inspect the system to find leaks. Many undetected leaks are "cold" leaks, which means that they only occur when the engine is cold and maybe only when it is running and cold. These are most often found at hose fittings, where clamps that perform well when warm are too loose during cold operation. When visually inspecting for leaks, don't forget to look inside the vehicle for leaks from the heater core or the hoses that attach to it. A complaint of a sweet smell in the cabin or steam on the windshield is your tip off that the heater core or heater hoses are leaking into the HVAC system.

If after visual inspection the leak is not found, applying pressure to the system with a pressure tester is the next order of business. When applying pressure, do not exceed the cap maximum pressure by more than 2 pounds. Do not forget to test the radiator cap. When a system does not hold pressure, it is not uncommon for coolant levels to look low because the coolant gets pushed out or into the recovery bottle and not drawn back into the system. This is because a vacuum does not occur, as it cools, to draw the coolant back into the radiator. This can also be your leak as the coolant expansion may be more than the recovery tank can hold. Keep in mind too that any leak that keeps the system from holding pressure can cause this.

Internal engine leakage is probably the worst case scenario in coolant leaks. Many times a persistent coolant loss is caused by a leaking head gasket or crack in an engine casting. It is important to diagnose this type of problem as early as possible because of other potential damage that can occur. Many leaks are between the combustion chamber and the cooling system. These, if undetected, can result in piston damage (since coolant does not compress). Radiator or heater core damage is common.

Most cooling systems run 13–18 psi; cranking pressure will easily be more than 100 psi. It does not take much to blow the tanks off of late-model plastic radiators. Internal leaks that aren't exposed to combustion pressures can fill up the crankcase with coolant and cause bearing damage. Internal leaks can be spotted by removing the dipstick or oil fill cap and looking for something like whitish brown whipped cream. These leaks are usually head gaskets or intake manifold gaskets on wet valley, V-type engines. In situations where a combustion leak is suspected but not readily apparent, you can often find them with a combustion leakage test that uses a colored liquid that changes color when exposed to exhaust gases in the cooling system.

The cooling system must help to maintain consistent temperature. An engine is a heat pump, but, because it does not have very high thermal efficiency, it makes more heat than can be turned into power. That heat is carried away to be exchanged with ambient air by the cooling system. The water pump moves the coolant through the engine and maintains adequate pressure in the block to keep bubbles from forming on castings that could create hot spots. The thermostat is a controlled restriction that helps the engine warm up by virtually stopping coolant flow to the radiator until a prescribed temperature when it opens. The thermostat may open and close during engine operation depending on ambient temperature and engine load. The radiator and heater core are air-to-water heat exchangers. The heater core is used to provide warm air inside the cabin and the radiator is used to remove extra heat from the engine. Late-model cars have very carefully chosen radiators to help them achieve operating temperature quickly. It does not take very much restriction of either airflow or coolant flow to cause a drop in radiator efficiency that results in overheating when under loads. The radiator's efficiency is tested by measuring the difference between inlet and outlet temperatures. A good rule of thumb is that a 40-degree drop is normal. The radiator may be restricted if that number is significantly higher or lower. Restriction can be in the form or debris collected in the radiator fins or deposits inside the tubes slowing or restricting coolant flow.

## Task C.4      Inspect and replace engine cooling and heater system hoses and pipes.

Check cooling system hoses for loose clamps, leaks, and damage. Hoses with cracks, abrasions, bulges, swelling, or that crunch when you squeeze them must be replaced.

Take care when removing hoses to avoid damage to radiator and heater core fittings. If the hose is to be replaced, it is wise to make a longitudinal cut in the hose and peel it off of the fitting. If the hose is good and will be reused, gently loosen it with a tool designed for it.

There are many types of belts, including V-belts, serpentine belts, and scalloped and wrapped construction type belts.

## Task C.5      Inspect, test, and replace radiator, pressure cap, coolant recovery system, and water pump.

There are three basic types of radiator caps: the constant pressure type, pressure vent type, and the closed system type. The closed system-type radiator cap is the one found on today's cars. The others are found on older model vehicles. The constant pressure type has a lower seal or pressure valve that is held closed until the coolant gets hot enough to build enough pressure to open the valve within the preset pressure range. The pressure vent-type cap is similar to the constant pressure type. However, it has a vacuum-release valve that is opened by a weight and is kept open to the atmosphere until the pressure is great enough to move the weight and close the valve. This prevents atmospheric pressure from entering into the radiator. Like the constant pressure cap, this cap opens to release pressure when it builds to the specified amount.

The closed system type works in the same way as the constant pressure cap, except it is designed to keep the radiator full at all times. When the specified pressure is reached, some coolant is released into the recovery tank. When there is a vacuum in the radiator (caused by less coolant), the vacuum is used to pull coolant from the recovery tank.

The vacuum valve in a radiator cap is designed to reduce cooling system vacuum caused by the contraction as a hot system cools. A sticking vacuum valve in the radiator cap may cause a vacuum in the system and collapse the upper radiator hose after the engine is shut off.

## Task C.6     Inspect, test, and replace thermostat, bypass, and housing.

The thermostat regulates minimum engine temperature by limiting the amount of coolant flow to the radiator during engine warmup and in cold weather operation. Computer-controlled vehicles will remain in a cold fuel strategy if the engine does not reach full operating temperature.

When the thermostat is closed, bypass hoses route water throughout the engine to insure even temperature throughout the engine until it reaches a temperature at which the radiator takes over those duties. Bypass hoses are often overlooked during hose replacement procedures and are often the source of hard-to-locate leaks.

Most thermostats have a housing that attaches them to the engine and facilitates the attachment of a radiator hose. These housings have been made of cast iron, aluminum, stamped steel, and plastic. They are often the site of small leaks that eventually corrode the housing, creating bigger leaks. When inspecting them, look for pitting or warpage. It used to be a common practice to file the housing. This was the cause of many thermostat failures because the thermostat usually fits in a recess and the housing provides clamping force. If the recess is in the housing, the thermostat may be clamped too tightly. This can cause it to distort and to drag or stick resulting in overheat or slow warmup. So before you try to save your customer a few bucks and wind up buying an engine, check this area carefully.

## Task C.7     Identify, inspect, and recover coolant; flush and refill system with proper coolant.

Cooling system service is a topic approached differently by technicians in different repair venues. We will offer information that should be generic to all repair technicians. ASE workshop participants must agree on the content of each question. Manufacturer-specific items will not appear in the test unless they are considered industry standard. It is very important to keep this in mind when taking the test.

When testing coolant, there are many methods available to reach the same end. We want to know the protection levels of the coolant for freezing, boiling, pH, corrosion protection, and in some vehicles, nitrites. Let us take a look at each area.

Freezing and boiling protection are linked, for the most part. Most all manufacturers agree that a mixture of 50% water to 50% coolant provides the best of both worlds in this area and the best component protection. All manufacturers will also agree that you should be sure to use the correct coolant in the vehicle without mixing coolant types or changing to one not designed for the vehicle.

pH is a measurement of the acidity or alkaline qualities of the coolant. As coolant becomes older, it drops toward the acid end of the pH scale. Most Asian vehicles aim for around 7–9 and most American and European vehicles aim for 8–9.5 on the scale. Low pH readings can be caused by deteriorated antifreeze condition or a water-heavy blend since water is more acidic than a coolant mix. Very high numbers can be caused by over-adding antifreeze or corrosion packages during service.

In normal use, vehicles with high output ignition systems, particularly DIS systems, will cause the coolant to become electrically charged, promoting debris in the system to adhere to metal parts and cause radiator restriction. This can really only be corrected by replacing the antifreeze or reversing the charge in the system, which some ionizing coolant recovery systems can do.

Corrosion protection is added when servicing the cooling system with recovery/recycling equipment and is in the antifreeze to begin with. This is a difficult area to test and a bone of contention with manufacturers who do not support coolant recycling.

The last area that will become more critical as more diesel vehicles enter the consumer market is nitrites. When out of balance, they cause small bubbles to collect on castings while the engine is running. These bubbles act like little cutters over time and carve into the casting. The vibration inherent in the diesel combustion process has been known to cause bubbles in the system to create leaks in cylinder walls. There are test strips available that detect the level of nitrites. Currently, only a couple of manufacturers have any specification for them for their light-duty diesel trucks.

**Task C.8**    **Inspect, test, and replace fan (both electrical and mechanical), fan clutch, fan belts, fan shroud, and air dams.**

Inspect serpentine belts for missing ribs, wear on outside edges, cracks closer than two inches apart, and excessive glazing. Any of these require immediate replacement. V-belts with cracks or excessive glazing on the surfaces require replacement. Many vehicles are equipped with plastic tensioner and idler pulleys. It is very important to inspect these for wear. They will allow belt tension to drop or even damage the belt when they wear. Belt tensioners have positioning marks that will indicate if the belt has reached a point at which it is not under proper tension.

When testing electric cooling fans, it is very important to have the wiring diagram in front of you before making any tests. Many cooling fans simply use a thermal switch to turn them on and a relay to bypass that switch when the air conditioner is turned on. More sophisticated systems use the coolant temperature sensor and the powertrain control module to operate the fan at varying speeds for different loads. They will receive an A/C command from the control head or one of the switches under the hood. The PCM may even vary speed based on system pressure. This multitude of variation means that questions on the test will have to present you with a wiring schematic or be general.

The mechanical fan systems used on air conditioned vehicles almost always incorporate a fan clutch, so it is a good idea to know that fan clutches come in two varieties. The first and most common is the thermostatic clutch that locks up the fan more as the engine temperature rises. The other type is the centrifugal fan clutch that tries to maintain a consistent speed regardless of engine speed. The two most common fan clutch failures are no fan clutch engagement at all or a completely locked up clutch. The first will result in poor air conditioner efficiency and overheating in traffic. The second will result in a complaint that the vehicle sounds like an airplane and that there is a loss of power as it can take up to an additional 17 horsepower to drive one of those multi-blade fans in a fixed mode at highway speeds. Check fan clutches for hydraulic leaks and looseness on the shaft, and fans for damaged blades. Many of the plastic fan blades in use will develop cracks that can cause the blade to break and hit the radiator core.

The purpose of a fan shroud is to allow a round fan to create a low-pressure zone behind the entire radiator core. The purpose of an air dam is to create a high-pressure zone in front of the radiator. Both components encourage airflow through the radiator under different circumstances. The fan shroud supports good airflow at low speeds and the air dam supports good airflow at highway speeds. This would lead us to keep in mind that a vehicle that overheats at highway speeds may have a problem with the air dam and a vehicle that overheats in traffic could have a problem with the fan shroud. The most common problems are that they are broken or missing.

**Task C.9**    **Inspect, test, and replace heater coolant control valve (manual, vacuum, and electrical types) and auxiliary coolant pump.**

The heater control valve (sometimes called the water flow valve) controls the flow of coolant into the heater core from the engine. In a closed position, the valve allows no flow of hot coolant to the heater core, keeping it cool. In an open position, the valve allows heated coolant to circulate through the heater core, maximizing heater efficiency.

Cable-operated valves are controlled directly from the heater control lever on the dashboard. Thermostatically controlled valves have a liquid-filled capillary tube located in the discharge airstream off the heater core. This tube senses air temperature, and the valve modulates the flow of water to maintain a constant temperature, regardless of engine speed or temperature.

Many heater valves utilized on today's cars are vacuum operated. These valves are normally located in the heater hose line or mounted directly in the engine block. When a vacuum signal reaches the valve, a diaphragm inside the valve is raised, either opening or closing the valve against an opposing spring. When the temperature selection on the dashboard is changed, vacuum to the valve is vented and the valve returns to its original position. Vacuum-actuated heater control valves are either normally open or normally closed designs. Some vehicles do not use a heater control valve; rather, a heater door controls how much heat is released into the passenger compartment from the heater core.

With cable-operated control valves, check the cable for sticking, slipping (loose mounting bracket), or misadjustment. With valves that are vacuum operated, there should be no vacuum to the valve when the heater is on (except for those that are normally closed and need vacuum to open).

On late-model vehicles, heater control valves are typically made of plastic for corrosion resistance and light weight. These valves feature few internal working parts and no external working parts. With the reduced weight of these valves, external mounting brackets are not required.

If the manual coolant control valve is incorrectly adjusted and does not close completely, hot coolant will be allowed to pass through the heater core when the A/C is on. If the coolant control valve does not open completely, it will not allow enough hot coolant to pass through the heater core to properly heat the air in the heat mode.

Rust discoloration at the heater control valve may indicate a malfunctioning valve. A defective heater control valve may result in a poor cooling condition by allowing hot coolant to enter the heater core. The heater control valve is used to control the flow of heated coolant through the heater core. The hot coolant is often used to "temper" the air for temperature control and to help maintain a desired in-car relative humidity. Heater coolant control valves may be electronically, mechanically, or vacuum controlled.

## Task C.10    Inspect, flush, and replace heater core.

Like the radiator, heater core tanks, tubes, and fins can become clogged over time by rust, scale, and mineral deposits circulated by the coolant. Heater core failures are generally caused by leakage or clogging. Feel the heater inlet and outlet hoses while the engine is idling and warm with the heater temperature control on hot. If the hose downstream of the heater valve does not feel hot, the valve is not opening.

If the heater core appears to be plugged, the inlet hose may feel hot up to the core but the outlet hose remains cool. Reverse flushing the core with a power flusher may open up the blockage, but usually the core has to be removed for cleaning or replacement. Air pockets in the heater core can also interfere with proper coolant circulation. Air pockets form when the coolant level is low or when the cooling system is not properly filled after draining.

A gurgling noise in the heater core may be caused by a low coolant level in the cooling system or a restricted heater core. A low coolant level will allow too much air in the cooling system. The excessive air in the system will mix with the coolant and create the gurgling noise. A restricted heater core will also cause a gurgling noise from the coolant passing through the restricted area.

# D.  Operating Systems and Related Controls Diagnosis and Repair (17 Questions)

## Task D.1    Electrical (9 Questions)

## Task D.1.1  Diagnose the cause of failures in the electrical control system of heating, ventilating, and A/C systems; determine needed repairs.

Electrical problems can be classified as one of three types: shorts, opens, and high-resistance problems. All of these have different effects on the system and their locations are found differently. A short is no more than an unwanted parallel circuit. Sometimes the short is a connection to the power side of another circuit; other times it is a connection to ground. In either case, a short will increase current flow in the circuit.

This typically will cause a fuse or circuit breaker to blow. An open is caused by a break or disconnected wire or component. No current will flow through the circuit. Unwanted high-resistance problems reduce the current flow through the circuit. They also reduce the amount of voltage available to the circuit's components. The most common high-resistance problem is a bad or loose ground.

An open circuit in the electronic door actuator circuit or a blown fuse would cause the temperature blend door actuator to be inoperative. An open circuit in the electronic temperature door actuator to the automatic temperature control circuit may cause an inaccurate temperature blend door position, but the door will move.

## Task D.1.2     Inspect, test, repair, and replace AC-heater blower motors, resistors, switches, relay/modules, wiring, and protection devices.

The blower motor/fan assembly is located in the evaporator housing. Its purpose is to increase airflow in the passenger compartment. The blower, which is basically the same type as those used in heater systems, draws warm air from the passenger compartment; forces it over the coils and fins of the evaporator; and blows the cooled, cleaned, and dehumidified air into the passenger compartment. The blower motor is controlled by a fan switch.

An open circuit at the blower switch ground or at the terminal causes the blower to be completely inoperative. An open circuit in the resistor or resistor terminals will cause the blower motor to operate at a speed not selected or not at all. An open at the switch terminal of the resistor will not allow the blower motor to operate at any speed selected by the switch. Keep in mind that some vehicles are equipped with blower fans that operate continuously even when off is selected.

A loose ground wire is often the cause of a no blower operating condition. A blower motor is used to force air through the heater core and air conditioner evaporator core. To function properly, the blower motor requires a complete electrical circuit. It is generally connected to the negative side of the battery via the frame of the vehicle and to the positive side of the battery via insulated wires, speed resistors, the switch, the fuse or circuit breaker, and the ignition switch.

## Task D.1.3     Inspect, test, repair, and replace A/C compressor clutch coil, relay/modules, wiring, sensors, switches, diodes, and protection devices.

Most late-model A/C systems have several controls intended to protect the compressor during extreme conditions like stop-and-go traffic, and to provide extra power when needed. Nearly all R-134A systems incorporate some way to measure both excessive high-side pressure and low-side pressure for either compressor cycling or to shut the system down if a major loss of refrigerant occurs. Here, again, look for generic questions that will test your understanding of how these could affect system performance.

Relays are used in A/C systems to provide the PCM or other low-amperage switches with a means to control functions like cooling fans, compressor clutch engagement, and/or wide-open throttle cutout used to drop out the compressor for additional power under load.

Some older vehicles used thermal-limiting fuses or switches that would shut down the compressor in the event of an overheat. These required technician intervention to make the system work again and were pretty much replaced by electronics by the early 1980s. This information is somewhat redundant to Task D.1.5.

## Task D.1.4     Inspect, test, repair, replace, and adjust A/C-related engine control systems.

Carbureted vehicles used electric solenoids to help the engine maintain an idle with the A/C compressor engaged. Most late-model vehicles use an A/C demand circuit to the PCM that, when activated, causes an adjusted idle strategy to perform this function. Driving conditions such as wide-open throttle and engine overheat may result in the PCM shutting down the AC system until the condition returns to normal.

## Task D.1.5     Inspect, test, repair, replace, and adjust load-sensitive A/C compressor cutoff systems.

Some vehicles have a power steering cutoff switch to disengage the air-conditioning compressor when the power steering requires maximum effort. Load-sensitive electrical switches include the low-pressure switch, high-pressure switch, pressure cycling, and power steering. Not all of these switches, however, are used on all vehicles. They are, for the most part, used to provide additional engine power when maximum power is required. The pressure switches are used to prevent compressor or component damage in the event of extremely high or low system pressures.

Some compressor clutch circuits contain a thermal limiter switch that senses compressor surface temperature. Some A/C compressor clutch circuits also contain a low-pressure and high-pressure cutoff switch.

## Task D.1.6    Inspect, test, repair, and replace engine cooling/condenser fan motors, relays/modules, switches, sensors, wiring, and protection devices.

A blown cooling fan fuse will obviously cause the cooling fans to be inoperative at all times. If the cooling fan operates only when A/C pressure is high, suspect an open A/C head pressure switch. If the system is equipped with high- and low-speed fan operations, an open circuit in one of the circuits may cause fan operation in only one speed.

Some cooling fan systems use a two-fan system; they have a low-speed fan and a high-speed fan. An electric cooling fan usually is wired through a relay because of the amount of current drawn to run an electric fan.

## Task D.1.7    Inspect, test, adjust, repair, and replace electric actuator motors, relays/modules, switches, sensors, wiring, and protection devices.

Electric actuator motors are used in ventilation systems to control the opening and positioning of various doors in the system. They are resistive and allow the HVAC control unit to "know" the position of each door.

Some actuator motors are calibrated automatically in the self-diagnostic mode, whereas other actuator motors must be calibrated manually. These actuators should only require calibration after motor replacement or misadjustment. Diagnostic trouble codes indicate fault in a specific component.

The electrical functions and circuits are different in each vehicle. Here as in all other areas where wiring, sensors, switches etc. are included, the question will have to provide you with a schematic or strategy to know what the system is supposed to be doing. Since ASE does not ask theory-based questions, the scenario must be a real-world work situation. Review the HVAC wiring diagrams of a couple of car lines you routinely service to make yourself familiar with the features of the system.

## Task D.1.8    Inspect, test, service, or replace heating, ventilating, and A/C control panel assemblies.

The negative battery cable must be disconnected and the technician must wait a specified length of time before performing the A/C control panel removal. Self-diagnostic tests may indicate a defective A/C control panel. The refrigeration system does not require discharging before A/C panel removal.

## Task D.2    Vacuum/Mechanical (3 Questions)

## Task D.2.1    Diagnose the cause of failures in the vacuum and mechanical switches and controls of the heating, ventilating, and A/C systems; determine needed repairs.

The purpose of the system is two-fold. It is used to house the heater core and the air conditioner evaporator and to direct the selected supply air through these components into the passenger compartment of the vehicle. The supply air selected can be either fresh (outside) or recirculated air, depending on the system mode. After the air is heated or cooled, it is delivered to the floor outlet, dash panel outlets, or the defrost outlets.

There are two basic duct systems employed. In the stacked core reheat system, the basic control is in the water valve. For maximum cold air, the water valve is completely closed. All air enters the vehicle compartment through the heater core. The access door, which is activated by a cable, controls only fresh or recirculated air.

Recirculated air is used during maximum cold operation; the air-conditioning unit is not operative and the evaporator will not be cold. The evaporator-only operation is used in the max air or maximum cold position. As the control level inside the car is moved, it controls the water valve by means of a vacuum or a cable to control the amount of hot water entering the heater core and the temperature of the air at the unit outlet.

In blend air reheat systems during heater-only operation, the air-conditioning unit is shut off, and the evaporator performs no function in air distribution or temperature control. During maximum air or extreme cold air, the air-conditioning system operates, the evaporator is cold, and the blend air door damper is completely closed. Only conditioned air enters the car.

As the control lever is moved in the vehicle from max air toward heat with the air conditioner on, the blend air door is moving. In maximum cold, it is completely shut. On maximum hot, it is completely open. The water valve on this unit is a vacuum on/off unit to regulate water flow. Normal position would be open. This type of blend air system is extremely popular and can be used with or without a water valve.

To check the proper functioning of the ductwork, move the temperature control lever to see if any change occurs. If it does not, shut off the air conditioner and turn on the heater. Move the temperature control arm again to see if any change occurs. If not, check the cable and the flap door connected to the temperature control lever. You might be able to reach under the dash to reconnect the cable or free a stuck flap.

If no substantial airflow is coming out of the registers, check the fuses in the blower circuit. Remove the fan switch and test it. Check the blower motor by hot-wiring it directly to the battery with jumper cables.

These are examples of very common system designs. There are many different designs so keep an open mind and expect that you will receive details about the type of system in question. Remember that you are being tested to see if you can work out a problem, not on your memorization of specifications. If specs are required, they will be given to you.

## Task D.2.2    Inspect, test, service, or replace heating, ventilating, and A/C control panel assemblies.

The control panel assembly controls the compressor, heater valve, and plenum door. The control panel assembly has no control over the coolant temperature. The purpose of the control panel is to provide operator input for the air-conditioning and heating system. Some control panels have provisions for displaying in-car and outside air temperatures. A microprocessor may be located in the control panel to provide input data to the programmer, based on operator-selected conditions.

The air conditioner (A/C) compressor operates with the control panel selector in the defrost position. The A/C compressor does not operate with the control panel selector in the vent position.

## Task D.2.3    Inspect, test, adjust, and replace heating, ventilating, and A/C control cables and linkages.

If the cable housing clamp is loose at the control head (panel) end, the valve may not function. If the cable were rusted in the housing, the control would not move freely.

A no-heat condition in a Bowden cable-controlled heater control valve system would not be caused by a binding cable holding the valve open. If the valve was held open, there would not be a no-heat condition. The no-heat condition could be caused by a kink in the cable holding the valve closed.

## Task D.2.4    Inspect, test, and replace heating, ventilating, and A/C vacuum actuators (diaphragms/motors) and hoses.

Transferring heated air from the heater core to passenger compartment heater and defroster outlets is the job of the heater and defroster ducts. The ducts are typically part of a large plastic shell that connects to the necessary inside and outside vents. This ductwork also has mounting points for the evaporator and heater core assemblies. Contained inside the duct are also the doors required to direct air to the floor, dash, and/or windshield. Sometimes the duct is connected directly to the vents; other times hoses are used.

A split, broken, or disconnected vacuum hose may affect the recirculation door operation. A clogged or blocked restrictor will not affect the recirculation door operation. A restricted or open check valve and cracked or broken vacuum tank may affect recirculation door operation. Vacuum motors move the various doors to direct the airflow to the proper vents. Vacuum air doors are controlled by a selector switch in the control head (panel). This unit directs the vacuum signal to the proper motor. If the selector switch is defective, vacuum will not be supplied to the proper door. If there is a leak in the vacuum system, the door will not move properly due to no- or a low-vacuum signal. A vacuum leak anywhere in the control circuit will affect the operation of the entire system.

When testing a vacuum-actuated door, the vacuum gauge reading should remain steady for at least one minute. If the gauge reading drops slowly, the actuator is leaking and should be replaced.

## Task D.2.5    Identify, inspect, test, and replace heating, ventilating, and A/C vacuum reservoir, check valve, and restrictors.

A leaking panel door actuator could cause the air discharge to switch from the panel to the floor ducts at any time. A leaking temperature blend door actuator also causes the system to change temperature at any time. A leaking intake manifold gasket may affect all the vacuum-operated mode doors because this condition causes low-source vacuum. A leaking check valve will not trap the vacuum in the reserve tank when climbing a long hill. This action may cause the temperature blend door to move to the warm air position and the air discharge to switch to the floor ducts.

Vacuum reservoirs are used to supply vacuum during periods of wide-open throttle (WOT) when the manifold vacuum is very low. Most vacuum reservoirs are connected to the vacuum source through a check valve.

## Task D.2.6    Inspect, test, adjust, repair, or replace heating, ventilating, and A/C ducts, doors, and outlets.

When the outside recirculation door is in position A, only outside air is allowed to enter the air conditioner heater case. When the outside recirculation door is in position B, only inside air is allowed to enter the air conditioner heater case. If the outside recirculation door is stuck in position A, outside air is drawn into the air conditioner heater case and there is no leakage of in-car air past this door.

All the air doors in an air conditioner heater case should move freely. When the mix position is selected on the control panel, the heater/defroster door should open about halfway.

## Task D.3    Automatic and Semiautomatic Heating, Ventilating, and A/C Systems (5 Questions)

## Task D.3.1    Diagnose temperature control system problems; determine needed repairs.

Temperature control systems for air conditioners usually are connected with heater controls. Most heater and air-conditioning systems use the same plenum chamber for air distribution. Two types of air-conditioning controls are used: manual/semiautomatic and automatic.

Air conditioner manual/semiautomatic temperature controls (MTC and SATC) operate in a manner similar to heater controls. Depending on the control setting, doors are opened and closed to direct airflow. The amount of cooling is controlled manually through the use of control settings and blower speed.

An automatic or electronic temperature control system maintains a specific temperature automatically inside the passenger compartment. To maintain a selected temperature, heat sensors send signals to a computer unit that controls compressor, heater valve, blower, and plenum door operation. A typical electronic control system might contain a coolant temperature sensor, in-car temperature sensor, outside temperature sensor, high-side temperature switch, low-side temperature switch, low-pressure switch, vehicle speed sensor throttle position sensor, sun-load sensor, and power steering cutout switch.

The control panel is found in the instrument panel at a convenient location for both driver and front-seat passenger access. Three types of control panels may be found: manual, pushbutton, or touch pad. All serve the same purpose. They provide operator input control for the air-conditioning and heating system. Some control panels have features that other panels do not, such as provisions to display in-car and outside air temperature in degrees.

Usually, a microprocessor is located in the control head to input data to the programmer, based on operator-selected conditions. When the ignition switch is turned off, a memory circuit remembers the previous setting. These conditions are restored the next time the ignition switch is turned on. If the battery is disconnected, however, the memory circuit is cleared and must be reprogrammed.

Many automotive electronic temperature control systems have self-diagnostic test provisions in which an on-board microprocessor-controlled subsystem displays a code. This code (number, letter, or alphanumeric) is displayed to tell the technician the cause of the malfunction. Some systems also display a code to indicate which computer detected the malfunction. Manufacturers' specifications must be followed to identify the malfunction display codes, since they differ from car to car.

If the in-car temperature sensor is defective, it may be sending a temperature signal that is colder than requested. This would cause the A/C temperature to be reduced to compensate for the cooler reading and thereby raise the in-car temperature. A sticking temperature blend door will cause warm air to be mixed with cool air and result in the in-car temperature being lower than the driver-selected temperature.

## Task D.3.2    Diagnose blower system problems; determine needed repairs.

The blower motor is usually located in the heater housing assembly. It ensures that air is circulated through the system. Its speed is controlled by a multiposition switch in the control panel. The switch works in connection with a resistor block that is usually located on the heater housing.

On some vehicles, when the engine is running, the blower motor is in constant operation at low speed. On automatic temperature control systems, the blower motor is activated only when the engine reaches a predetermined temperature. The blower motor circuit is protected by a fuse located in the fuse panel. The fuse rating is usually 20 to 30 amperes.

The blower motor resistor block is used to control the blower motor speed. The typical resistor block is composed of three or four wire resistors in series with the blower motor, which control its voltage and current. The speed of the motor is determined by the control panel switch, which puts the resistors in series. Increasing the resistance in the system slows the blower speed.

If the blower does not operate, use a test light to make sure there is voltage on both sides of the fuse. Then, check to see if current is arriving at the motor. On cars where the blower motor is behind the inner fender shell, hunt out the wiring and check for current. If the blower motor is getting current, the problem is either a burned out blower motor or a bad ground on the motor. In situations where no current is available at the motor, backtrack to check for an open resistor. Check also for burned or corroded connections in the blower relays or bulkhead connectors. An open blower motor ground would cause the motor not to run at all. If the correct amount of voltage is available to the motor and the ground is good, suspect a bad motor.

## Task D.3.3    Diagnose air distribution system problems; determine needed repairs.

With the temperature selector in the cool position, the incoming air should be the same temperature as that of the outside air. A stuck blend door would cause the incoming air to be heated with the temperature selector in the cool position.

With VENT selected on the control panel, air is directed to the vents as normal. When the accelerator pedal is more than 3/4 of the way depressed, the air is directed to the defrost vents. This would not be caused by a stuck air door. This could be caused by a bad check valve between the vacuum reservoir tank and the supply vacuum.

## Task D.3.4    Diagnose compressor clutch control system; determine needed repairs.

If the A/C pressure cutoff switch is stuck open, the compressor clutch will not engage. If the compressor clutch coil is shorted to ground, the compressor clutch will not engage. If the compressor clutch coil has an opening in its windings, the compressor clutch will not engage. If the compressor clutch coil positive wire is shorted to a 12-volt source, the compressor clutch will not disengage.

If a compressor clutch circuit fuse is open, there is no voltage supplied to the circuit. The PCM grounds the clutch control relay winding. An open circuit would make it impossible for the PCM to ground a relay winding.

## Task D.3.5    Inspect, test, adjust or replace climate and blower control sensors.

The in-vehicle sensor is a thermistor. As temperature increases, resistance decreases; as temperature decreases, resistance increases. Resistance is at a maximum when the temperature in the vehicle is cold. As the temperature of the in-vehicle sensor increases, the sensor resistance should decrease.

A sun-load sensor usually cannot be adjusted. The outside air temperature sensor does not affect the sun-load sensor calibration. Therefore, if the sun-load sensor does not function, its diagnosis would be simple. Electronic control of blower speed by pulse width modulation requires the use of speed feedback sensors.

### Task D.3.6    Inspect, test, adjust, and replace door actuator(s).

On vehicles with a blend air reheat system and during heater-only operation, the air-conditioning unit is shut off, and the evaporator performs no function in air distribution or temperature control. During maximum air or extreme cold air, the air-conditioning system operates, the evaporator is cold, and the blend air door damper is completely closed. Only conditioned air enters the car.

As the control lever is moved in the vehicle from max air toward heat with the air conditioner on, the blend air door is moving. In maximum cold, it is completely shut. On maximum hot, it is completely open. The water valve on this unit is a vacuum on/off unit to regulate water flow. Normal position would be open. This type of blend air system is extremely popular and can be used with or without a water valve.

To check the proper functioning of the ductwork, move the temperature control lever to see if any change occurs. If it does not, shut off the air conditioner and turn on the heater. Move the temperature control arm again to see if any change occurs. If not, check the cable and the flap door connected to the temperature control lever. You might be able to reach under the dash to reconnect the cable or free a stuck flap.

If no substantial airflow is coming out of the registers, check the fuses in the blower circuit. Remove the fan switch and test it. Check the blower motor by hot-wiring it directly to the battery with jumper cables.

In some systems, the temperature blend door is controlled by an electric door actuator that is controlled by an A/C computer. In some A/C systems, the temperature blend door is controlled by a vacuum actuator that may be called a power servo in which a varying vacuum is supplied. It operates the solenoid or solenoids controlled by the A/C computer.

### Task D.3.7    Inspect, test, and replace heater water valve and controls.

If the coolant control valve was defective, the hose between it and the heater core would not be hot. The indication is that hot coolant is passing the control valve but not exiting the heater core. Therefore, the heater core may be clogged and should be cleaned or replaced.

Heater control valves are usually mechanically, electrically, or vacuum controlled. When the maximum A/C mode is selected, the valve is usually closed to maximize the cooling effect.

### Task D.3.8    Inspect, test, and replace electric and vacuum motors, solenoids, and switches.

The process of testing vacuum and electric motors in an automatic temperature control (ATC) system is the same as for non-ATC vehicles with the exception that you may receive help in diagnosing them from the ATC system itself by way of trouble codes or through functional tests.

### Task D.3.9    Inspect, test, and replace automatic temperature control (ATC) control panel and/or climate control computer (microprocessor/programmer).

As with many other subjects, this one is made complex by the different manufacturers' approach to the same system. Remember the things that are always true of ATC or ETC systems:

1. They all have a control head or computer that runs the system.

2. The control unit has inputs to monitor components for their safety and temperature inputs.

3. The control unit has outputs that control relays, solenoids, motors and communicate with other on-board control units.

These are the basics of any electronic climate control system.

You should know that it may be possible to perform testing of the system by either the control head or in other systems by use of a scan tool. You will be provided with any specifications you might need and an explanation of the system in question.

## Task D.3.10 Check and adjust calibration of automatic temperature control (ATC) system.

The specified voltage drop across computer ground wires usually is 0.1 V. Anything above that would indicate a bad ground or open wire causing a high resistance. With the low voltages used on computer-controlled climate systems, a voltage drop that is higher could cause a problem with the system.

To adjust an electronic actuator on a computer-controlled air-conditioning system, you have to place the air conditioner controls in the specified mode, move the slide link by hand to the specified door position, and rotate the actuator to the specified position.

The ATC system may be "out of calibration" due to a defective in-car sensor or a restricted aspirator. There are three main automatic temperature control (ATC) component groups: the sensors and control head (panel) resistors, the programmer, and the control servo motors. If any of these components are defective or out of specifications, overall performance of the ATC system may suffer.

# E. Refrigerant Recovery, Recycling, and Handling (6 Questions)

## Task E.1   Maintain and verify correct operation of certified equipment.

Air-conditioning (A/C) recovery/recycling equipment must have a UL approval and SAE J1991 approval. Refrigerant oils for R-12 and R-134A must not be mixed. The refrigerant container specified by the recovery/recycling equipment manufacturer must be used in this type of equipment to ensure that the container has proper capacity and valving.

Refrigerant recovery/recycle equipment must have UL approval and meet SAE standard J1991. The federal Clean Air Act (CAA), Section 609, established the law that anyone who performs a service involving refrigerant in an automobile air-conditioning system must be properly certified. In addition, all equipment used in this service must have a label certifying that it is UL approved and meets SAE standard J1991.

## Task E.2   Identify and recover A/C system refrigerant.

A refrigerant identifier should be used to determine the type of refrigerant and if it includes contamination. . Most identifiers list the percentages of R134a, R12 and air, as well as if the system contains any flammable types of refrigerant. A technician should never connect the shops equipment up to a vehicle that has shown to have a mixture of refrigerants or any flammable refrigerants. Most refrigerant identifiers will not detect if sealer is installed in an A/C system. There are some sealer detection kits available to test for refrigerant that might contain a sealer. Sealants can damage recovery/recycling equipment so exercise caution to prevent this damage. Recovery is accomplished by pumping the refrigerant out of the system and into a special container. Some machines also will recycle the refrigerant by filtering it, removing oil and moisture. If, after the recovery process, the low-side gauge rises above 0 psi, there is still some refrigerant remaining in the system. The system should be evacuated until there is no pressure left in the system.

The container used for recycled refrigerant R-12 is stamped DOT4BA. The storage containers for recycled R-12 refrigerant must qualify for DOT CFR Title 49. Such a container is stamped DOT4BA or DOT4BW for identification.

## Task E.3   Recycle or properly dispose of refrigerant.

If the moisture warning light is on during the recycling process, the refrigerant contains excessive moisture and the filter/cartridge in the recovery/recycling equipment must be changed. After recovery, many refrigerant recovery systems display the amount of refrigerant and oil recovered from the system.

A multipass system may not complete all recycling stages before storing the refrigerant. UL approval is required of all recovery/recycle equipment. SAE standard J1991 is also required. There are two types of UL-approved recycle equipment; single pass and multipass. A single-pass system will not remove all contaminants. A multipass system will, however, clean and dry refrigerant to standards for reuse before dispensing it.

## Task E.4    Label and store refrigerant.

Refrigerant storage containers must be filled to 60 percent of their gross weight rating. Refrigerant storage containers must be evacuated to 17 in. Hg (43.9 kPa absolute) before refrigerant is placed in the container.

When storing recovered refrigerant, use a DOT4BW cylinder. Never store refrigerant in anything else. A DOT 39 cylinder is not safe to use for storing recovered refrigerant.

## Task E.5    Test recycled refrigerant for noncondensable gases.

When checking a refrigerant container for noncondensable gases (air), store the container out of the presence of sunlight at 65°F (18°C) for 12 hours, and place a thermometer 4 inches (102 mm) from the container surface. If the container pressure is within specifications, the refrigerant is ready for use.

Before testing recycled refrigerant for noncondensable gases, place the portable refrigerant cylinder in a relatively cool area, out of direct sunlight, for no less than 12 hours. The ambient temperature may be as low as 65°F (18°C). Compare the temperature and pressure reading with an appropriate pressure limit chart.

## Task E.6    Follow federal and local guidelines for retrofit procedures.

Converting a R-12 system to use R-134A can be done if all federal and local regulations are met. Typically the R-12 system must be thoroughly purged and flushed to remove all traces of R-12. Because R-134A operates at higher pressures than R-12, new refrigerant hoses, designed for those higher pressures, may be required. Different systems require that different components be changed as well.

After the conversion has been completed, the system must be labeled as an R-134A system. All standard or required service procedures (such as reclaiming) must be followed during conversion.

# 5 Sample Test for Practice

## Sample Test

*Please note the letter and number in parentheses following each question. They match the task in Section 4 that discusses the relevant subject matter. You may want to refer to the overview using the cross-referencing key to help with questions posing problems for you.*

1. An A/C compressor is noisy when the clutch is engaged but the noise goes away when the clutch is disengaged. Which of these is the MOST-Likely cause?
   - A. Faulty compressor clutch pulley bearing
   - B. Low refrigerant charge
   - C. Internal compressor damage
   - D. Overtightened drive belt                                                    (B.1.6)

2. While conducting a pressure test on an A/C system, the ambient temperature is 80°F, the low-side gauge reads low (8 psi), and the high side also reads low (85 psi). Which of these could cause these readings?
   - A. Low refrigerant level
   - B. Normal operation
   - C. Bad compressor
   - D. A high-side restriction                                                     (A.1)

3. An A/C system is being tested with pressure gauges. The high-side pressure is low, and the low-side pressure is high. Which of these is the LEAST-Likely cause of these pressures?
   - A. Expansion valve stuck open
   - B. Restriction in the high side of the system
   - C. The compressor's head gasket is leaking
   - D. The POA or STV is stuck closed                                              (A.1)

4. Which of the following oils is the standard oil used by OEM's in R-134A systems?
   - A. Mineral oil
   - B. CCOT
   - C. Ester oil
   - D. PAG                                                                         (A.9)

5. Technician A says the lines before and after the receiver/dryer should be the same temperature. Technician B says the lines before and after the evaporator should be the same temperature. Who is right?
   - A. A only
   - B. B only
   - C. Both A and B
   - D. Neither A nor B                                                             (A.3)

6. All of the following statements are true **EXCEPT:**
   A. refrigerant leaves the compressor as a high-pressure, high-temperature vapor.
   B. moisture and contaminants are removed by the receiver/drier, where the moisture is stored until it is needed.
   C. the expansion valve controls the flow of refrigerant into the evaporator.
   D. warm air passing through the evaporator causes heat to be absorbed into the refrigerant.                                                                                    (A.3)

7. Technician A says if an air-conditioning system has too much refrigerant oil, the performance of the system will suffer. Technician B says poor system performance can be caused by an overcharge of refrigerant in the system. Who is right?
   A. A only
   B. B only
   C. Both A and B
   D. Neither A nor B                                                                           (A.1)

8. A customer complains that the air-conditioning system cools satisfactorily during the early morning or late evening, but does not cool during the hot part of the day. During testing, the low-side gauge starts out reading normal, then drops into a vacuum. Technician A says that ice could be forming on the expansion valve. Technician B says that the drier could be saturated with moisture. Who is right?
   A. A only
   B. B only
   C. Both A and B
   D. Neither A nor B                                                                           (A.3)

9. A customer complains about engine coolant loss. The cooling system is pressurized at 15 psi (103 kPa) for 15 minutes. There is no visible sign of coolant leaks in the engine or passenger compartments, but the pressure on the tester gauge decreases to 5 psi (34 kPa). This problem could be caused by any of the following defects **EXCEPT:**
   A. a leaking heater core.
   B. a leaking internal transmission cooler.
   C. a leaking head gasket.
   D. a cracked cylinder head.                                                                  (C.3)

10. The vacuum valve in the radiator cap is stuck closed. The result of this problem could be:
    A. an upper radiator hose collapsed after the engine was shut off.
    B. excessive cooling system pressure at normal engine temperature.
    C. engine overheating when operating under a heavy load.
    D. engine overheating during extended idle periods.                                         (C.4)

11. After the recovery of an A/C system with a recovery machine, which of these statements is true of the systems condition?
    A. All refrigerant and oil have been removed from the system.
    B. The system is ready for normal use.
    C. All refrigerant has been removed from the system.
    D. Humidity has been removed from the refrigerant in the system.                            (A.6)

12. An air-conditioning system equipped with an accumulator is a:
    A. thermostatic expansion valve system.
    B. fixed orifice tube system.
    C. receiver/drier system.
    D. throttling suction valve system.                                                         (A.1)

13. Technician A says the R-12 refrigeration system does not need to be depressurized to replace the high-pressure relief valve. Technician B says if replacing the high-pressure relief valve on a retrofitted R-134A system, it is necessary to replace it with one that has the same relief pressure specification as the original relief valve. Who is right?
    A. A only
    B. B only
    C. Both A and B
    D. Neither A nor B                                                    (B.2.11)

14. A customer says that he thinks that an alternative refrigerant was added to his car when the A/C was serviced last. Technician A says the system must be purged of all refrigerant before it can be evaluated. Technician B says that a refrigerant identifier should be used to determine the type of refrigerant. Who is right?
    A. A only
    B. B only
    C. Both A and B
    D. Neither A nor B                                                    (A.7)

15. Technician A says that some A/C systems control evaporator pressure by varying the displacement of the compressor. Technician B says some systems regulate the flow of refrigerant into the evaporator with a thermal expansion valve (TXV). Who is right?
    A. A only
    B. B only
    C. Both A and B
    D. Neither A nor B                                                    (B.2.8)

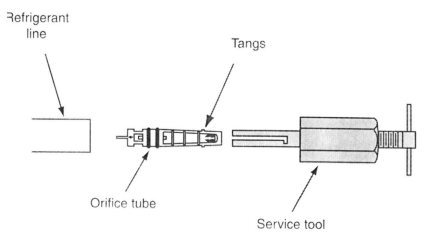

16. All of the following statements are true about using the tool shown in the figure to remove an orifice tube assembly **EXCEPT:**
    A. pour a small amount of refrigerant oil on top of the orifice tube to lubricate the O-rings.
    B. rotate the T-handle to engage the notch in the tool in the orifice tube tangs.
    C. rotate the T-handle to remove the orifice tube from the evaporator inlet pipe.
    D. hold the T-handle and rotate the outer sleeve on the tool to remove the orifice tube. (B.2.6)

17. Technician A says if the suction line to the compressor is covered with thick frost, this might indicate that the expansion valve is flooding the evaporator. Technician B says frost on a line between the condenser outlet and the expansion valve indicates a restriction to the flow of refrigerant. Who is right?
   A. A only
   B. B only
   C. Both A and B
   D. Neither A nor B                                                                 (A.8)

18. The coolant level in the coolant recovery container is normal when the engine is cold. This level becomes much higher than normal after the vehicle has been driven for 45 minutes. Technician A says some of the radiator tubes may be restricted. Technician B says the radiator cap may be defective. Who is right?
   A. A only
   B. B only
   C. Both A and B
   D. Neither A nor B                                                                 (C.5)

19. The high-side pressure is excessively high for a thermal expansion valve (TXV) system. There is frosting on the condenser near the outlet of the condenser before the receiver/drier. Technician A says the cause may be a refrigerant overcharge. Technician B says there may be a restriction in the condenser. Who is right?
   A. A only
   B. B only
   C. Both A and B
   D. Neither A nor B                                                                 (B.2.3)

20. Cabin air filters remove dust and pollen from the:
   A. recirculated air.
   B. outside air.
   C. all AC air sources.
   D. heated air only.                                                                (A.10)

21. All of the following statements about refrigerant system tools are true **EXCEPT:**
   A. separate gauges and other refrigerant handling equipment should be used for different types of refrigerant.
   B. manifold gauge sets for R-134A can be identified by unique service fitting connections.
   C. R-134A service hoses are much stronger because of the high system pressures.
   D. proper identification of service equipment and hoses is important to prevent cross-contamination of refrigerant.                                              (A.5)

22. There is oil residue around the high-pressure relief valve and the AC refrigerant charge has been determined to be low. Technician A says the cause could be restricted air passages through the condenser. Technician B says the refrigerant system may have been overcharged. Who is right?
   A. A only
   B. B only
   C. Both A and B
   D. Neither A nor B                                                                 (B.1.1)

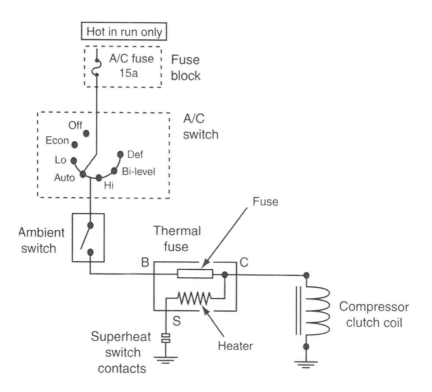

23. In the A/C control system shown in the figure, the ignition switch is on, and the A/C switch is in the AUTO position. The ambient temperature is 75°F (24°C), and the compressor clutch is inoperative. There are 12 volts at Terminals B and C on the thermal fuse. The cause of the inoperative compressor clutch could be:
   A.  an open thermal fuse.
   B.  an open compressor clutch coil.
   C.  a defective superheat switch.
   D.  a defective thermal fuse heater.                                  (B.1.2)

24. A check of the A/C system with a manifold and gauge set indicates that, after 15 minutes of operation, the low side gauge indicates a pressure below zero. Technician A says the cause could be a faulty compressor cycling switch. Technician B says that the fixed orifice tube may be clogged. Who is right?
   A.  A only
   B.  B only
   C.  Both A and B
   D.  Neither A nor B                                                   (A.2)

25. A stuck-open engine cooling system thermostat could cause all of the symptoms **EXCEPT:**
   A.  poor heater operation.
   B.  below-normal engine operating temperature.
   C.  increased fuel consumption.
   D.  engine overheating.                                              (C.6)

26. A car is equipped with an A/C compressor driven by a V-type belt. As the compressor cycles, the belt squeals briefly. A check with a belt tension gauge indicates the belt is correctly adjusted. Technician A says the belt may be too narrow for the pulley. Technician B says that the system may be overcharged with refrigerant. Who is right?
    A. A only
    B. B only
    C. Both A and B
    D. Neither A nor B                                                                (B.1.3)

27. Technician A says a replacement compressor should have the same type of clutch and pulley as the old compressor. Technician B says the mounting brackets and other fasteners on a replacement compressor should be identical to those on the old compressor. Who is right?
    A. A only
    B. B only
    C. Both A and B
    D. Neither A nor B                                                                (B.1.6)

28. Technician A says that the oil level of some compressors may be checked with the use of a dipstick. Technician B says the oil must be drained from some compressors and measured in a graduated container. Who is right?
    A. A only
    B. B only
    C. Both A and B
    D. Neither A nor B                                                                (A.9)

29. An A/C system operates normally for about 10 miles of driving and then stops cooling. The thermal expansion valve (TXV) is frosted. If the A/C system is shut off for five minutes and turned on again, it blows cool air for another 10 mi. (16 km). The MOST-Likely cause of this problem is:
    A. a defective thermal expansion valve (TXV) capillary tube.
    B. a refrigerant overcharge.
    C. a restricted condenser refrigerant passage.
    D. moisture contamination in the refrigeration system.                            (B.2.5)

30. A customer complains that the air-conditioning system cools satisfactorily during the early morning or late evening, but does not cool during the hot part of the day. During testing, the low-side gauge starts out reading normal, then drops into a vacuum. Technician A says that ice could be forming on the expansion valve. Technician B says that the drier could be saturated with moisture. Who is right?
    A. A only
    B. B only
    C. Both A and B
    D. Neither A nor B                                                                (A.3)

31. The airflow through the AC condenser is blocked by debris. Which symptom will the system have?
    A. Poor cooling
    B. Freezing of the evaporator
    C. Poor airflow from the dash outlets
    D. Loss of refrigerant                                                            (B.2.2)

32. Frost is forming on one of the condenser tubes near the bottom of the condenser. This problem could be caused by:
    A. restricted airflow passages in the condenser.
    B. a refrigerant leak in the condenser.
    C. a restricted refrigerant passage in the condenser.
    D. a restricted fixed orifice tube.                                    (B.2.3)

33. When the A/C is in operation, the duct outlet temperature is higher than normal on a vehicle equipped with a normally closed vacuum heater control valve. Which of these might be the cause?
    A. A vacuum leak at the heater control valve
    B. Debris collected in the heater core inlet
    C. The heater control valve stuck open
    D. The heater control valve stuck closed                               (C.1)

34. Refer to the system shown in the diagram. Technician A says that there should be vacuum to the top hose on the floor/defroster valve actuator when heat is selected on the control panel. Technician B says that there should be vacuum at the panel door valve actuator when defrost is selected on the control panel. Who is right?
    A. A only
    B. B only
    C. Both A and B
    D. Neither A nor B                                                     (D.2.4)

35. The control Bowden cable of a heater control valve moves freely, but the valve does not respond. Technician A says the cable housing clamp may be loose at the control head (panel) end. Technician B says the cable may be rusted in the housing. Who is right?
    A. A only
    B. B only
    C. Both A and B
    D. Neither A nor B                                                     (D.2.3)

36. The low-side pressure is higher than normal. Cooling the expansion valve remote bulb lowers the low-side pressure. Technician A says the expansion valve is defective. Technician B says the remote bulb may be improperly secured to the evaporator outlet. Who is right?
    A. A only
    B. B only
    C. Both A and B
    D. Neither A nor B                                                     (B.2.5)

37. Technician A says a restricted orifice tube may cause higher-than-specified low-side pressure. Technician B says a restricted orifice tube may cause frosting of the orifice tube. Who is right?
    A. A only
    B. B only
    C. Both A and B
    D. Neither A nor B                                                     (B.2.6)

38. Technician A says the air conditioner may be charged through the high side with the system running as long as the refrigerant container is not inverted. Technician B says inverting the container causes low-pressure refrigerant vapor to be charged into the system. Who is right?
    A. A only
    B. B only
    C. Both A and B
    D. Neither A nor B                                                                    (A.8)

39. The inside of the windshield has an oily film, and the A/C cooling is inadequate. Technician A says this oily film may be caused by a plugged A/C heater case drain. Technician B says this oily film may be caused by a leak in the evaporator core. Who is right?
    A. A only
    B. B only
    C. Both A and B
    D. Neither A nor B                                                                    (B.2.7)

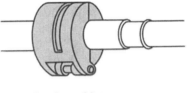

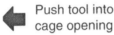
Push tool into
cage opening

40. When servicing refrigerant lines, the tool shown in the figure is used to:
    A. disconnect spring lock couplings.
    B. connect spring lock couplings.
    C. connect and disconnect spring lock couplings.
    D. crimp male barb-type fittings.                                                     (B.2.1)

41. Technician A says that the evaporator temperature regulator (ETR) and evaporator pressure regulators (EPRs) are devices used for evaporator temperature control in older systems. Technician B says that a suction throttling valve (STV) is a device used to control evaporator temperature on newer systems. Who is right?
    A. A only
    B. B only
    C. Both A and B
    D. Neither A nor B                                                                    (B.2.9)

42. Which of these would NOT be used to detect an A/C refrigerant leak on an R-134A system?
    A. Flame-type leak detector
    B. Halogen leak detector
    C. Dye and UV light
    D. Audible noise from the leak                                                        (A.4)

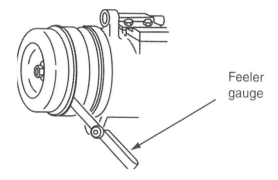

Feeler
gauge

43. The measurement in the figure exceeds the specification. Technician A says this could cause an intermittent scraping noise with the engine running and the compressor clutch engaged. Technician B says this could cause a slipping compressor clutch. Who is right?
    A. A only
    B. B only
    C. Both A and B
    D. Neither A nor B
    (B.1.4)

44. Technician A says a Schrader-type service valve is used to quickly and easily connect and disconnect refrigerant service equipment to the system. Technician B says when manifold gauge set hoses are connected to the Schrader-type service valves, a pin in the hose end should open the valve. Who is right?
    A. A only
    B. B only
    C. Both A and B
    D. Neither A nor B
    (B.2.10)

45. A rattling sound is coming from the area of the AC compressor when in operation. When the compressor is shut off, the noise stops. The AC system cools well and there are no indications of A/C system problems. Technician A says the noise could be caused by a loose compressor mount. Technician B says that the noise could be caused by broken fasteners of the compressor mounting. Who is right?
    A. A only
    B. B only
    C. Both A and B
    D. Neither A nor B
    (B.1.7)

46. Technician A says the desiccant in a receiver/drier absorbs moisture and it can become quickly contaminated if exposed to atmosphere. Technician B says orifice tube systems use an accumulator to store excess refrigerant and to filter and dry the refrigerant. Who is right?
    A. A only
    B. B only
    C. Both A and B
    D. Neither A nor B
    (B.2.4)

47. The inside of the windshield has a sticky film. Technician A says the engine coolant level should be checked. Technician B says the heater core may be leaking. Who is right?
    A. A only
    B. B only
    C. Both A and B
    D. Neither A nor B
    (C.2)

48. All the following statements about cooling system service are true **EXCEPT:**
    A.  when the cooling system pressure is increased, the boiling point is decreased.
    B.  if more antifreeze is added to the coolant, the boiling point is increased.
    C.  good quality ethylene glycol antifreeze contains antirust and corrosion inhibitors.
    D.  coolant solutions must be recovered, recycled, or handled as hazardous waste.          (C.7)

49. A heavy ice buildup is noted on the thermostatic expansion valve. The sight glass appears to be clear. Technician A says the thermostatic expansion valve is the coldest part in the system and the condition is normal. Technician B says this could be caused by a restriction in the receiver/drier. Who is right?
    A.  A only
    B.  B only
    C.  Both A and B
    D.  Neither A nor B                                                                         (A.3)

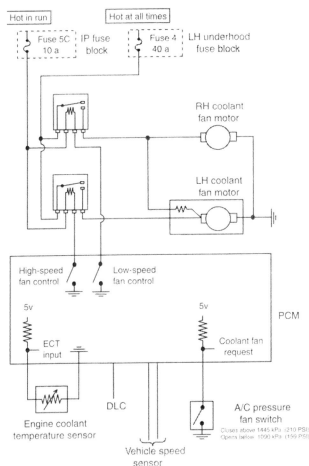

50. In the electric cooling fan circuit shown in the figure, the low-speed and high-speed fans do not operate when the engine is overheated. Both fans will run when the air conditioning is turned on. The cause of this problem could be:
    A.  a blown 10A, 5C fuse in the instrument panel fuse block.
    B.  a defective engine coolant temperature sensor.
    C.  a defective high-speed coolant fan relay.
    D.  a defective A/C pressure fan switch.                                                    (C.8)

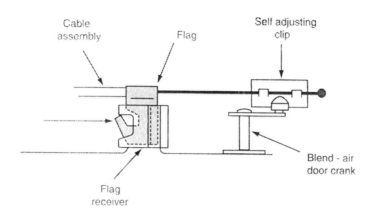

51. In the figure, the adjustment being performed is to the:
    A. air blend door.
    B. manual coolant control valve.
    C. ventilation door control rod.
    D. defroster door control rod.                                      (C.9)

52. Technician A says that a blower resistor is used to lower voltage to the blower motor to create different speeds. Technician B says that the resistance provided by the blower resistor is highest at the highest blower speed. Who is right?
    A. A only
    B. B only
    C. Both A and B
    D. Neither A nor B                                                  (D.1.2)

53. Technician A says a loose ground wire may be the cause of no blower operation. Technician B says no blower operation may be caused by a defective compressor clutch diode. Who is right?
    A. A only
    B. B only
    C. Both A and B
    D. Neither A nor B                                                  (D.1.2)

54. Technician A says a purpose of the compressor clutch diode is to prevent alternating current from the alternator entering the clutch coil. Technician B says a purpose of the diode is to prevent spikes of high voltage created by the clutch coil operation from damaging delicate electronic components. Who is right?
    A. A only
    B. B only
    C. Both A and B
    D. Neither A nor B                                                  (D.1.3)

55. Technician A says the ambient temperature switch senses outside air temperature and is designed to prevent compressor clutch engagement when air conditioning is not required because of low ambient temperatures. Technician B says the switch is in series with the compressor clutch electrical circuit and closes at about 37°F to prevent clutch engagement. Who is right?
    A. A only
    B. B only
    C. Both A and B
    D. Neither A nor B                                                  (D.3.5)

56. Technician A says that some vehicles have an rpm or throttle position sensor that disengages the A/C compressor during times of heavy load. Technician B says some vehicles have a power steering cutoff switch to disengage the A/C compressor when the power steering requires maximum effort. Who is right?
    A.  A only
    B.  B only
    C.  Both A and B
    D.  Neither A nor B                                                                        (D.1.5)

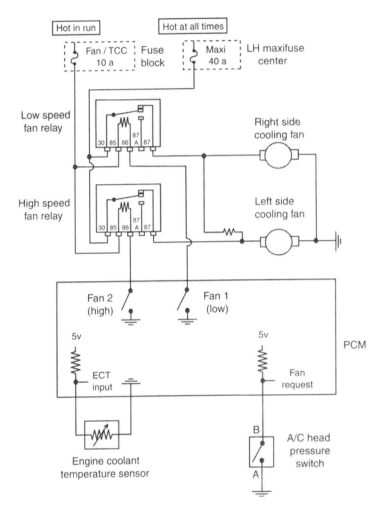

57. The cooling fan system shown in the figure, operates normally on high speed, but there is no low-speed fan operation under any operating condition. The cause of this problem could be:
    A.  an open circuit at pin 30 of the low-speed relay.
    B.  a blown 40 amp maxifuse in the LH maxifuse center.
    C.  a blown 10 amp cooling fan/TCC fuse in the fuse block.
    D.  an open circuit at Terminal B on the A/C head pressure switch.                          (D.1.6)

58. All of the following statements about computer-controlled A/C system actuator motors are true **EXCEPT:**
    A. some actuator motors are calibrated automatically in the self-diagnostic mode.
    B. A/C system and component problems do not produce diagnostic trouble codes.
    C. the actuator motor control rods must be calibrated manually on some systems.
    D. the actuator motor control rods should only require calibration after motor replacement or misadjustment. (D.1.7)

59. All of the following statements about A/C heater control panel service are true **EXCEPT:**
    A. the negative battery cable should be disconnected before A/C control panel removal.
    B. the refrigeration system must be discharged before the A/C control panel is removed.
    C. if the vehicle is air bag-equipped, wait the specified time period after disconnecting the negative battery cable.
    D. self-diagnostic tests may indicate a defective A/C control panel in a computer-controlled A/C system. (D.1.8)

60. None of the vacuum controls is operational with the engine running at idle speed. Technician A says the problem could be caused by a disconnected vacuum supply hose. Technician B says the manifold vacuum fitting may be blocked. Who is right?
    A. A only
    B. B only
    C. Both A and B
    D. Neither A nor B (D.2.1)

61. A typical control panel assembly controls all of the following **EXCEPT** the:
    A. compressor.
    B. heater valve.
    C. plenum door.
    D. coolant temperature. (D.2.2)

62. When diagnosing a computer-controlled A/C system, a diagnostic trouble code (DTC) is obtained indicating a fault in the temperature blend door actuator motor. Technician A says the first step in the repair procedure is to replace the temperature blend door actuator. Technician B says to disconnect the battery and test the system again. Who is right?
    A. A only
    B. B only
    C. Both A and B
    D. Neither A nor B (D.3.8)

63. A gurgling noise is heard inside the passenger compartment. The noise is coming from the A/C heater case. Technician A says the heater core coolant passages may be restricted. Technician B says the coolant level in the cooling system may be low. Who is right?
    A. A only
    B. B only
    C. Both A and B
    D. Neither A nor B (C.10)

64. An A/C system has a vacuum reservoir, a check valve, and a vacuum-operated mode door actuator including the blend-air door. While operating in the A/C mode and climbing a steep hill with the throttle nearly wide open, the air discharge temperature gradually becomes warm, and the air discharge switches from the panel to the floor ducts. The A/C system operates normally under all other conditions. The cause of this problem could be:
    A. a leaking panel door vacuum actuator.
    B. a defective vacuum reservoir check valve.
    C. a leaking blend-air door vacuum actuator.
    D. a leaking intake manifold gasket.                                    (D.2.5)

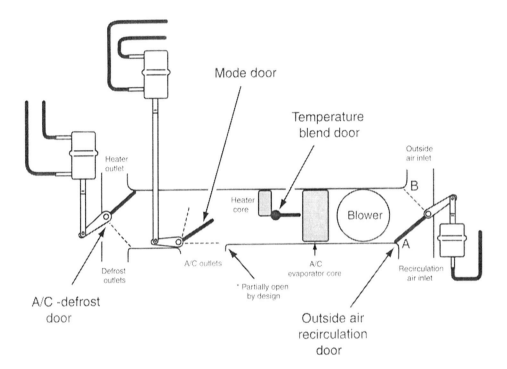

65. In the figure, the air-recirculation/fresh air door is stuck in position "A." Technician A says under this condition, outside air is drawn into the A/C-heater case. Technician B says under this condition, there would be no heater operation. Who is right?
    A. A only
    B. B only
    C. Both A and B
    D. Neither A nor B                                                      (D.2.6)

66. On a semiautomatic A/C system, the temperature control is set at 70°F (21°C), and the in-car temperature is 80°F (27°C) after driving the car for 1 hour. The refrigerant system pressures are normal. Technician A says the in-car sensor may be defective. Technician B says the temperature blend door may be sticking. Who is right?
    A. A only
    B. B only
    C. Both A and B
    D. Neither A nor B                                                      (D.3.1)

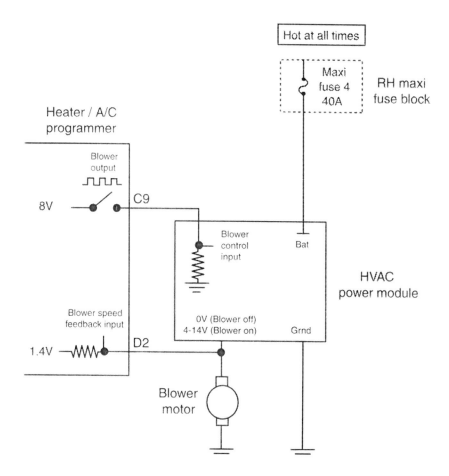

67. In the figure, the blower motor will not run at any speed setting. When the fan speed control on the A/C controls is changed from off to low speed, there is voltage at terminal D2 on the heater/A/C programmer. The cause of this problem could be:
    A. a defective blower motor.
    B. a blown 40A fuse #4.
    C. a defective HVAC power module.
    D. an open circuit at programmer terminal D2.                                    (D.3.2)

68. In the vent position and with the temperature selector in the cool position, the incoming air is heated. What could be the problem?
    A. The engine cooling fan is inoperative.
    B. The engine coolant level is low.
    C. The heater core's coolant passages are restricted.
    D. An air blend door is stuck.                                                  (D.3.3)

69. Which of the following is NOT recommended while retrofitting an R-12 system to R-134A?
    A. Install R-134A conversion fittings on the high and low side service ports.
    B. Install conversion labels next to the R-12 label.
    C. Recharge the system with R-134A to approximately 80 percent of the original R-12 charge.
    D. Use a refrigerant identifier to make sure the system only contains pure R-12 before recovery.                                                               (E.6)

70. Technician A says that restricted refrigerant passages in the evaporator may cause frosting of the evaporator outlet pipe. Technician B says restricted refrigerant passages in the evaporator may cause much higher than specified low-side pressures. Who is right?
    A. A only
    B. B only
    C. Both A and B
    D. Neither A nor B                                                                    (B.2.6)

71. The compressor clutch will not disengage when the A/C control is switched to off. The problem could be:
    A. the A/C pressure cutoff switch is stuck open.
    B. the compressor clutch coil is shorted to ground.
    C. the low-pressure switch has an open wire.
    D. the compressor coil feed circuit is shorted to voltage.                            (D.3.4)

72. The heater water valve is being checked. The hose between the control valve and heater core is hot, and the outlet hose from the heater core is much cooler. Technician A says the coolant control valve may be defective and should be cleaned or replaced. Technician B says the heater core may be clogged and should be cleaned or replaced. Who is right?
    A. A only
    B. B only
    C. Both A and B
    D. Neither A nor B                                                                    (D.3.7)

73. All of the following statements about A/C recovery/recycling equipment are true **EXCEPT:**
    A. the equipment label must indicate UL approval.
    B. the equipment label must indicate SAE J1991 approval.
    C. any size and type of refrigerant storage container over 10 lb.(4.8 kg) may be used in this equipment.
    D. R-12 and R-134A refrigerants or refrigerant oils must not be mixed in the recovery/recycling process.                                                          (E.1)

74. After the recovery process, the low-side pressure increases above zero after five minutes. This condition indicates:
    A. there is still some refrigerant in the system.
    B. there is excessive oil in the refrigerant system.
    C. the refrigerant system is leaking.
    D. there is excessive moisture in the refrigerant system.                             (E.2)

75. A recovery machine shuts down during the recovery process. An indicator light is signaling FULL. Technician A says the car's system was overcharged and has reached the normal refrigerant level. Technician B says the recovery cylinder may be 80% full. Who is right?
    A. A only
    B. B only
    C. Both A and B
    D. Neither A nor B                                                                    (E.3)

76. During refrigerant recovery, a cylinder is considered full when it reaches what capacity:
    A. 50 percent.
    B. 60 percent.
    C. 70 percent.
    D. 80 percent.                                                                        (E.4)

77. When checking a refrigerant container for noncondensable gases,:
   A. the container should be stored at 80°F (27°C) for six hours before the test.
   B. the container may be stored near a shop window.
   C. a thermometer should be placed against the container surface.
   D. if the pressure is lower than specified, the refrigerant is ready for use.          (E.5)

78. A refrigerant identifier can tell if there is:
   A. air in the refrigerant system.
   B. an excessive refrigerant charge.
   C. excessive oil in the refrigerant system.
   D. a low refrigerant charge.                                                           (A.7)

# 6 Additional Test Questions for Practice

## Additional Test Questions

*Please note the letter and number in parentheses following each question. They match the task in Section 4 that discusses the relevant subject matter. You may want to refer to the overview using the cross-referencing key to help with questions posing problems for you.*

1. Refer to the figure. The blower motor speed is believed to be slow at all speeds. It is determined that the blower will not reach full speed when on high. The voltmeter is reading 8 volts in the high-speed switch position Technician A says the cause could be low source voltage. Technician B says the problem could be a high resistance at the blower motor ground. Who is right?
   A. A only
   B. B only
   C. Both A and B
   D. Neither A nor B                                                        (D.1.2)

2. While connecting a manifold gauge set to an R-12 A/C system, Technician A says the high-pressure hose should connect to the line between the evaporator and compressor. Technician B says the low-pressure hose should connect to the service port on the outlet of the compressor. Who is right?
   A. A only
   B. B only
   C. Both A and B
   D. Neither A nor B                                                        (A.1)

3. When charging an empty air-conditioning system through the high-pressure side, all of the following conditions would be true **EXCEPT:**
   A. the system must be off.
   B. the system should be under a vacuum.
   C. the refrigerant drum should be inverted for liquid charging.
   D. no refrigerant will enter the low side of the system.                  (A.8)

4. Technician A says a restricted orifice tube could be caused by debris from a failing compressor. Technician B says an orifice tube restriction would cause a low-side pressure that is considerably lower than specified. Who is right?
   A. A only
   B. B only
   C. Both A and B
   D. Neither A nor B                                                        (B.2.6)

5. While discussing the conversion of an R-12 system to an R-134A system, Technician A says the R-12 system must be thoroughly purged and flushed to remove all traces of R-12. Technician B says that all oil should be removed from the system and replaced with the appropriate R-134A-compatible oil. Who is right?
    A.  A only
    B.  B only
    C.  Both A and B
    D.  Neither A nor B                                               (E.6)

6. Technician A says a slipping fan clutch will reduce ram airflow. Technician B says reduced airflow across the condenser results in low suction pressure. Who is right?
    A.  A only
    B.  B only
    C.  Both A and B
    D.  Neither A nor B                                               (B.2.2)

7. Technician A says while retrofitting an A/C system, it is recommended to replace all O-ring seals. Technician B says the accumulation of oily residue around an A/C line connection is an indication of a leak in the system. Who is right?
    A.  A only
    B.  B only
    C.  Both A and B
    D.  Neither A nor B                                               (B.2.2)

8. The air passages through an A/C condenser are severely restricted. Technician A says this may cause refrigerant discharge from the high-pressure relief valve. Technician B says this may cause the high-side pressure and low-side pressure to be higher than normal. Who is right?
    A.  A only
    B.  B only
    C.  Both A and B
    D.  Neither A nor B                                               (B.2.2)

9. Technician A says when the engine is running with the thermostat closed, the coolant is directed through a bypass system to ensure coolant circulation. Technician B says some systems use the heater core as a thermostat bypass. Who is right?
    A.  A only
    B.  B only
    C.  Both A and B
    D.  Neither A nor B                                               (C.6)

10. Technician A says a vacuum pump is used during the recovery process to remove the last trace of refrigerant from the system. Technician B says any air or moisture not evacuated from the air-conditioning system reduces the system's efficiency. Who is right?
    A.  A only
    B.  B only
    C.  Both A and B
    D.  Neither A nor B                                               (A.6)

11. Technician A says that some high-pressure relief devices are spring-loaded and self-resetting. Technician B says that some pressure relief devices are made of fusible metal and are not self-resetting. Who is right?
    A.  A only
    B.  B only
    C.  Both A and B
    D.  Neither A nor B                                                        (B.2.11)

12. Technician A says that the pressure cycling switch should cause the compressor to cycle OFF when the low-side pressure goes below about 25 psi. Technician B says that the pressure cycling switch will cause the compressor to cycle ON when the low-side pressure reaches about 47 psi. Who is right?
    A.  A only
    B.  B only
    C.  Both A and B
    D.  Neither A nor B                                                        (B.2.9)

13. There is a noticeable noise coming from the engine compartment when the A/C is selected. Technician A says this could be caused by a loose compressor mount. Technician B says this could be caused by the system being empty of refrigerant. Who is right?
    A.  A only
    B.  B only
    C.  Both A and B
    D.  Neither A nor B                                                        (B.1.7)

14. When retrofitting a system to R-134A, Technician A says that it is important to verify that the existing compressor is compatible with the new refrigerant. Technician B says that all system O-rings and gaskets compatible with mineral oil should be replaced. Who is right?
    A.  A only
    B.  B only
    C.  Both A and B
    D.  Neither A nor B                                                        (E.6)

15. Technician A says a leaking heater core is often the cause of windshield fogging Technician B says that cleaning a clogged evaporator/heater case drain tube may eliminate a windshield fogging problem. Who is right?
    A.  A only
    B.  B only
    C.  Both A and B
    D.  Neither A nor B                                                        (C.2)

16. Technician A says coolant is considered a hazardous waste and must be recovered. Technician B says the coolant may be reused by recycling and adding an additive. Who is right?
    A.  A only
    B.  B only
    C.  Both A and B
    D.  Neither A nor B                                                        (C.7)

17. An A/C compressor cycles off when the high-side pressure builds. Technician A says that the high-pressure cutoff switch is opening the circuit to the compressor clutch. Technician B says that the pressure release valve is opening the circuit to the compressor clutch. Who is right?
    A.  A only
    B.  B only
    C.  Both A and B
    D.  Neither A nor B                                                    (B.2.9)

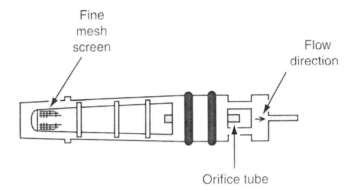

Fine
mesh
screen

Flow
direction

Orifice tube

18. The filter screen of an orifice tube is covered with aluminum particles. Which of these could be the cause?
    A.  A leaking desiccant bag
    B.  A plugged condenser
    C.  A damaged compressor
    D.  A damaged evaporator                                              (B.2.6)

19. When interpreting gauge readings, an elevated high-side reading may be caused by all of these **EXCEPT:**
    A.  refrigerant overcharge.
    B.  restricted airflow to the condenser.
    C.  poor airflow across the evaporator.
    D.  a slipping fan clutch.                                            (A.1)

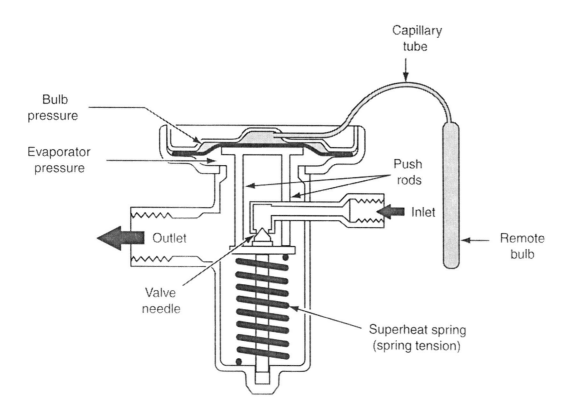

20. In the figure, the capillary tube is broken. What is the vehicle symptom that will likely occur as a result?
    A. The air conditioning will always blow at full cold.
    B. The air conditioning will not blow cold enough.
    C. It will become impossible to charge the air conditioning.
    D. It will not affect the air conditioning.                                    (B.2.5)

21. Technician A says that a loose air-conditioning belt can squeal whether or not the compressor clutch is engaged. Technician B says that a slipping serpentine belt can be caused by a defective automatic tensioner. Who is right?
    A. A only
    B. B only
    C. Both A and B
    D. Neither A nor B                                                             (B.1.3)

22. The high-side pressure on a system with a cycling clutch and pressure cycling switch is 245 psi, the low-side pressure hovers around 28 psi, and the ambient temperature is 98°F. What do these gauge readings indicate?
    A. The system is undercharged.
    B. The system is overcharged.
    C. The system is normal.
    D. The evaporator pressure regulator is bad.                                   (A.1)

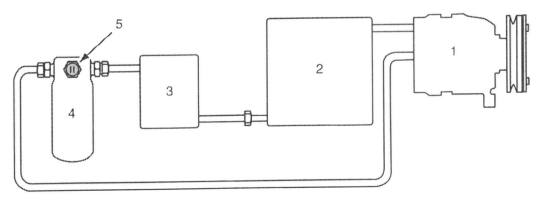

23. In the figure of the orifice tube-type refrigerant system, Component 5 could be all of these
    **EXCEPT:**
    A. low-pressure switch.
    B. high-pressure cutoff switch.
    C. cycling clutch switch.
    D. low refrigerant charge switch.                                    (B.1.1)

24. While testing a compressor clutch, a fused jumper wire is used to apply voltage directly from
    the battery to the clutch coil connector. This does not cause the compressor clutch to engage.
    The LEAST-Likely cause of this is:
    A. an open compressor clutch coil.
    B. a shorted compressor clutch coil.
    C. an open compressor clutch circuit ground.
    D. an open clutch coil feed circuit.                                  (D.3.4)

25. A vehicle that is stopped stalls when the steering wheel is turned to full lock only when the air
    conditioning is on. Technician A says that the vehicle may have a bad power steering pressure
    switch. Technician B says this could occur if the air-conditioning belt is loose. Who is right?
    A. A only
    B. B only
    C. Both A and B
    D. Neither A nor B                                                    (D.1.5)

26. An A/C system is being leak tested. Technician A carefully checks all system connections and
    says the presence of oil around the fitting of an air-conditioning line or hose is an indication of a
    possible refrigerant leak. Technician B uses a hand-held electronic leak detector and moves the
    hose slowly along all components of the refrigerant system. Who is right?
    A. A only
    B. B only
    C. Both A and B
    D. Neither A nor B                                                    (A.4)

27. A customer complains that the engine is overheating. While inspecting the vehicle when it is
    hot, it is found to have excess coolant in the coolant recovery container. After the engine cools,
    the level returns to normal. Technician A says that this could be caused by a restriction in the
    radiator. Technician B says that a blown head gasket could also cause this. Who is right?
    A. A only
    B. B only
    C. Both A and B
    D. Neither A nor B                                                    (C.3)

28. The engine overheats when the vehicle is parked, with the engine running. When the vehicle
    is driven at highway speeds, the engine operates at the normal temperature. Technician A
    says a missing or broken radiator fan shroud could be the cause. Technician B says a seized
    thermostatic fan clutch could be the cause. Who is right?
    A. A only
    B. B only
    C. Both A and B
    D. Neither A nor B                                                      (C.8)

29. While diagnosing the cause of a blown fuse in the blower motor circuit, Technician A says a
    short in the blower motor circuit could cause the fuse to blow. Technician B says an open field
    winding in the blower motor could have caused the fuse to blow. Who is right?
    A. A only
    B. B only
    C. Both A and B
    D. Neither A nor B                                                      (D.1.2)

30. The fins and air passages of an evaporator are heavily clogged. Technician A says the evaporator
    core must be removed from the case for proper cleaning. Technician B says there would not
    have been a problem if the evaporator case drain had been cleaned periodically. Who is right?
    A. A only
    B. B only
    C. Both A and B
    D. Neither A nor B                                                      (B.2.8)

31. A technician finds that an air-conditioning system has excessive high-side pressure. The
    low-side pressure is normal to high. A probable cause of this problem is:
    A. the condenser air passages are restricted.
    B. a leaking thermostatic expansion valve.
    C. an open bypass valve.
    D. a malfunctioning vapor control switch.                               (A.2)

32. While conducting an A/C system performance test, Technician A inserts a thermometer into the
    air duct at the center of the dash and monitors discharge temperature. Technician B conducts a
    visual inspection. Who performed a correct step in the process?
    A. A only
    B. B only
    C. Both A and B
    D. Neither A nor B                                                      (A.1)

33. In most automatic temperature control systems, Technician A says that the control panel
    controls compressor operation. Technician B says that some electronic control panels contain
    microprocessors. Who is right?
    A. A only
    B. B only
    C. Both A and B
    D. Neither A nor B                                                      (D.2.2)

34. Technician A says a high-side charging procedure should be completed with the engine running. Technician B says if liquid refrigerant enters the compressor, damage to the compressor may result. Who is right?
    A. A only
    B. B only
    C. Both A and B
    D. Neither A nor B                                              (A.8)

35. Technician A says that A/C cycling pressure switches are used to prevent evaporator icing on fixed-orifice systems. Technician B says that thermostatic deicing switches prevent icing on the condensers of systems that use thermostatic expansion valves. Who is right?
    A. A only
    B. B only
    C. Both A and B
    D. Neither A nor B                                              (B.1.1)

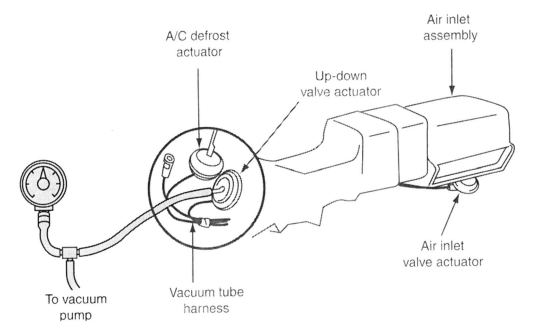

36. An air distribution actuator door vacuum motor circuit is being tested using the tool as shown. The tool is connected into the vacuum line going to the engine, and a zero vacuum reading is displayed. This would indicate:
    A. the air distribution actuator motor door is restricted.
    B. the line to the engine is plugged, disconnected, or kinked.
    C. the vacuum motor is defective.
    D. the motor should be operational.                            (D.2.1)

37. Evacuating a refrigerant system removes what from the system?
    A. Moisture
    B. Rust
    C. Dirt
    D. Desiccant particles                                         (A.6)

38. When the heating system delivers an insufficient amount of heat, Technician A says the problem could be a faulty engine thermostat. Technician B says the problem could be an inoperative cooling fan. Who is right?
    A.  A only
    B.  B only
    C.  Both A and B
    D.  Neither A nor B                                                                (C.1)

39. No heat is coming from the heater outlets. Technician A says the heater core may be restricted. Technician B says the engine cooling system thermostat may be stuck open. Who is right?
    A.  A only
    B.  B only
    C.  Both A and B
    D.  Neither A nor B                                                                (C.1)

40. Technician A says some compressor clutch circuits contain a thermal limiter switch that senses compressor surface temperature. Technician B says some A/C compressor clutch circuits have a low-pressure cutoff switch. Who is right?
    A.  A only
    B.  B only
    C.  Both A and B
    D.  Neither A nor B                                                                (D.1.5)

41. A car makes a hissing noise each time the A/C system and engine are turned off. Technician A says that the noise is caused by a refrigerant leak. Technician B says that the noise is caused by equalization of system pressures. Who is right?
    A.  A only
    B.  B only
    C.  Both A and B
    D.  Neither A nor B                                                                (A.3)

42. Technician A says a container of refrigerant oil must be kept closed when not in use to prevent the oil from absorbing moisture. Technician B says anytime an A/C system has been opened to atmosphere, the system needs to be evacuated after the repair. Who is right?
    A.  A only
    B.  B only
    C.  Both A and B
    D.  Neither A nor B                                                                (A.9)

43. While diagnosing higher-than-normal-system pressures, Technician A checks the condenser for dirt buildup in the condenser's fins. Technician B tests the engine for indications of an overheating condition. Who is right?
    A.  A only
    B.  B only
    C.  Both A and B
    D.  Neither A nor B                                                                (A.1)

44. Technician A says if the refrigerant is contaminated with moisture, chemical changes in the refrigerant can cause corrosion or ice may form internally. Technician B says if refrigerant oil is allowed to mix with the refrigerant, the system will not build up the necessary pressures to work efficiently. Who is right?
    A. A only
    B. B only
    C. Both A and B
    D. Neither A nor B                                                                    (A.7)

45. A receiver/drier is located between the condenser and the evaporator. Technician A says the receiver/drier should be changed if the outlet is colder than the inlet. Technician B says the receiver/drier should be changed if the refrigerant in the sight glass appears red. Who is right?
    A. A only
    B. B only
    C. Both A and B
    D. Neither A nor B                                                                   (B.2.4)

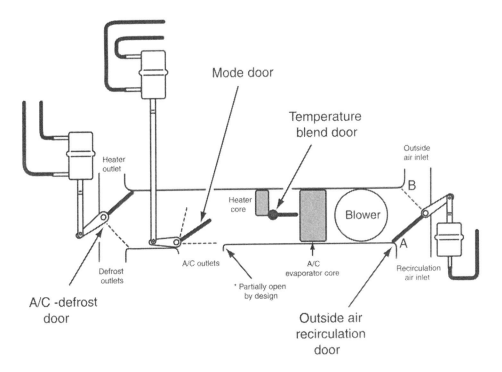

46. In the figure, Technician A says in order for the incoming air to be directed through the heater core, it must go through the evaporator core first. Technician B says the incoming air must go through the evaporator even if heat is selected. Who is right?
    A. A only
    B. B only
    C. Both A and B
    D. Neither A nor B                                                                   (D.2.4)

47. All of the following statements about refrigerant lines are true **EXCEPT:**
    A. suction lines are located between the outlet side of the evaporator and the inlet side or suction side of the compressor.
    B. suction lines carry the low-pressure, low-temperature refrigerant vapor to the compressor where it again is recycled through the system.
    C. suction lines can be distinguished from the discharge lines by touch; they are hot to the touch.
    D. the suction line is larger in diameter than the liquid line because refrigerant in a vapor state takes up more room than refrigerant in a liquid state.          (B.2.1)

48. All of these statements about refrigerant recovery are true **EXCEPT:**
    A. the recycling equipment must have shutoff valves within 12 inches of the hoses' service ends.
    B. the shutoff valves should be open when connecting the hoses to the vehicle's air-conditioning service fittings.
    C. recover the refrigerant from the vehicle and continue the process until the vehicle's system shows vacuum instead of pressure.
    D. turn off the recovery/recycling unit for at least five minutes, and then check the pressure of the system.          (A.5)

49. The oil required in most R-134A refrigerant systems is:
    A. a polyalkylene glycol (PAG) oil.
    B. a synthetic engine oil.
    C. a synthetic mineral oil.
    D. petroleum oil.          (A.9)

50. A vacuum actuator is being tested by applying 18 in. Hg. (40.5 kPa absolute) of vacuum to the vacuum actuator. Technician A says the vacuum gauge reading should remain steady for at least one minute. Technician B says if the gauge reading drops slowly, the actuator is leaking. Who is right?
    A. A only
    B. B only
    C. Both A and B
    D. Neither A nor B          (D.2.5)

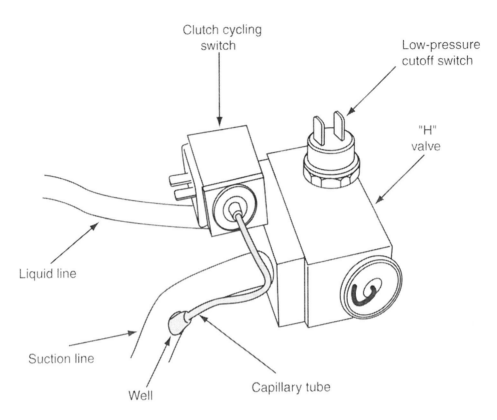

51. Technician A says that warming the sensing bulb of a capillary tube should cause the system pressures to change. Technician B says that cooling the above sensing bulb should cause the system pressures to change. Who is right?
    A.  A only
    B.  B only
    C.  Both A and B
    D.  Neither A nor B                                                      (B.2.5)

52. An air-conditioning clutch does not engage the compressor. A voltage measurement shows there is 12V at the clutch coil feed wire. Technician A says that the clutch coil could be bad. Technician B says that the clutch coil ground could be bad. Who is right?
    A.  A only
    B.  B only
    C.  Both A and B
    D.  Neither A nor B                                                      (B.1.2)

53. Which of the following statements about possible heating system problems is NOT true?
    A.  With the heater temperature control on hot, the water control valve should be open.
    B.  A sticking or disconnected blend door will cause the heat to only come out of the floor outlets.
    C.  Normally closed heater control valves that are vacuum-operated should have vacuum when the heater is on.
    D.  Air pockets in the heater core can interfere with proper coolant circulation.       (A.3)

54. Technician A says that a system with 1/4 inch service valves can only be serviced with R-12. Technician B says that if no labels are on the vehicle indicating that it is to be serviced with R-134A, then it must contain R-12. Who is right?
    A. A only
    B. B only
    C. Both A and B
    D. Neither A nor B                                                                    (A.5)

55. The blower motor does not operate on high speed. All other speeds are operational. Technician A says the blower ground wire may be loose or broken. Technician B says the high-speed relay may be defective. Who is right?
    A. A only
    B. B only
    C. Both A and B
    D. Neither A nor B                                                                    (D.1.7)

56. When checking an A/C system, the gauges read normal pressures but the evaporator's inlet and outlet tubes feel like they are near the same temperature. Measurement confirms that the evaporator outlet pipe is only 10°F colder than the inlet. Which of these conditions could cause this?
    A. A plugged evaporator
    B. A normal system condition
    C. Restricted orifice tube
    D. Leaking condenser                                                                  (A.3)

57. Refer to the figure of manual stem-type service valves. All of the following statements are true **EXCEPT:**
    A. when in the front seated position indicated as A, the compressor can be replaced.
    B. when in the back seated position indicated as B, the system can be operated normally.
    C. when in the mid position, the refrigerant can be recovered.
    D. when in the mid position, the system should not be operated.                       (A.9)

58. Technician A says that R-12 is sold in white containers marked with a DOT code. Technician B says that R-134A refrigerant is sold in light blue containers. Who is right?
    A. A only
    B. B only
    C. Both A and B
    D. Neither A nor B                                                                    (A.5)

59. When a fixed-orifice tube (FOT) cycling clutch A/C system is operating at 82°F (27.8°C) ambient temperature, the compressor clutch cycles several times per minute and the evaporator outlet line is warm. The high-side gauge shows lower-than-normal pressures. There is no frost on any of the A/C system components. The cause of this problem could be:
    A. a low refrigerant charge.
    B. a flooded evaporator.
    C. a restricted accumulator.
    D. an overcharge of refrigerant.                                                      (A.3)

60. A customer complains that the air-conditioning system cools satisfactorily during the early morning or late evening, but does not cool during the hot part of the day. During testing, the low-side gauge starts out reading normal, then drops into a vacuum. Technician A says that ice could be forming on the expansion valve. Technician B says that the drier could be saturated with moisture. Who is right?
    A.  A only
    B.  B only
    C.  Both A and B
    D.  Neither A nor B                                                                      (A.3)

61. To charge an air-conditioning system while it is running, the refrigerant should be added to:
    A.  the high side.
    B.  the low side.
    C.  both the high and low sides.
    D.  either the high or the low side.                                                     (A.10)

62. Failure to change the cabin air filter according to the maintenance schedule could result in:
    A.  excessive high-side pressure.
    B.  reduced airflow from the outlets.
    C.  a frozen evaporator.
    D.  a restricted evaporator drain.                                                       (A.10)

63. The customer complains that the A/C is not as cool as it used to be. The customer indicates that the A/C was serviced recently by another shop and it worked for a while. Technician A says the first thing to do is check the pressure with a manifold and gauge set. Technician B says the first thing to do is recover the refrigerant. Who is right?
    A.  A only
    B.  B only
    C.  Both A and B
    D.  Neither A nor B                                                                      (A.7)

64. Technician A says a thermal switch can be connected in series with the compressor clutch. Technician B says a thermal switch is usually mounted in the compressor. Who is right?
    A.  A only
    B.  B only
    C.  Both A and B
    D.  Neither A nor B                                                                      (B.1.2)

65. Technician A says some cooling fan systems have a low-speed fan and a high-speed fan. Technician B says an electric cooling fan is usually controlled through a relay. Who is right?
    A.  A only
    B.  B only
    C.  Both A and B
    D.  Neither A nor B                                                                      (D.1.6)

66. Technician A says that some computer-controlled A/C system actuator motors are automatically calibrated in the self-diagnostic mode. Technician B says that some computer-controlled A/C system actuator motors require manual calibration after replacement. Who is right?
    A.  A only
    B.  B only
    C.  Both A and B
    D.  Neither A nor B                                                                      (D.1.7)

67. Technician A says that the evaporator housing and water drain are integral parts of the evaporator housing assembly. Technician B says that a crack or break in the housing may reduce the cooling capacity of the air conditioner. Who is right?
    A. A only
    B. B only
    C. Both A and B
    D. Neither A nor B                                                                          (B.2.8)

68. A compressor clutch coil can be checked for shorts by measuring:
    A. current with an ammeter.
    B. voltage with a voltmeter.
    C. resistance with an ohmmeter.
    D. magnetic strength with a magnetometer.                                                    (B.1.4)

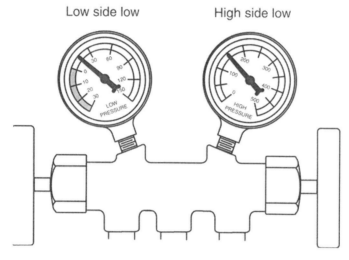

Low side low                          High side low

69. The gauge pressures shown in the figure occur on an R-12 thermal expansion valve (TXV), cycling clutch air-conditioning (A/C) system with an ambient temperature of 80°F (26.7°C). The cause of these readings could be:
    A. air in the refrigerant system.
    B. a defective A/C compressor.
    C. a low refrigerant charge.
    D. a stuck-open TXV.                                                                         (A.2)

70. Technician A says the type of oil used when a system is retrofitted should be listed on the retrofit label. Technician B says that all R-134a systems use the same oil. Who is right?
    A. A only
    B. B only
    C. Both A and B
    D. Neither A nor B                                                                          (A.9)

71. An oily residue is present on the fittings of the hose connected from the compressor to the condenser. Technician A says this residue may be caused by excessive oil in the refrigerant system. Technician B says this residue may be caused by a leak at the hose fittings. Who is right?
    A. A only
    B. B only
    C. Both A and B
    D. Neither A nor B                                                                          (A.4)

72. In an R-12, thermal expansion valve (TXV), cycling clutch A/C system with an ambient temperature of 80°F (27.8°C), the low-side pressure is 55 psi (345 kPa) and the high-side pressure is 260 psi (1,793 kPa). There is no indication of frosting on any of the refrigerant system components, and the discharge air is slightly cool. Technician A says the cause of this problem may be air in the refrigerant system. Technician B says the cause of this problem may be a restricted TXV. Who is right?
    A.  A only
    B.  B only
    C.  Both A and B
    D.  Neither A nor B                                                                    (A.2)

73. In an A/C system, if the low-side pressure is lower than normal, which of the following is the LEAST-Likely cause?
    A.  A faulty metering device
    B.  Poor airflow across the evaporator
    C.  A restriction in the low side of the system
    D.  The system is overcharged with refrigerant                                         (A.1)

# 7 Appendices

## Answers to the Test Questions for the Sample Test Section 5

| | | | | | | | |
|---|---|---|---|---|---|---|---|
| 1. | C | 21. | C | 41. | A | 61. | D |
| 2. | A | 22. | C | 42. | A | 62. | D |
| 3. | B | 23. | B | 43. | C | 63. | C |
| 4. | D | 24. | C | 44. | C | 64. | B |
| 5. | A | 25. | D | 45. | C | 65. | A |
| 6. | B | 26. | C | 46. | C | 66. | C |
| 7. | C | 27. | C | 47. | C | 67. | A |
| 8. | C | 28. | B | 48. | A | 68. | D |
| 9. | A | 29. | D | 49. | D | 69. | B |
| 10. | A | 30. | C | 50. | B | 70. | A |
| 11. | C | 31. | A | 51. | A | 71. | D |
| 12. | B | 32. | C | 52. | A | 72. | B |
| 13. | D | 33. | C | 53. | A | 73. | C |
| 14. | B | 34. | A | 54. | B | 74. | A |
| 15. | C | 35. | A | 55. | A | 75. | B |
| 16. | C | 36. | B | 56. | C | 76. | D |
| 17. | C | 37. | B | 57. | A | 77. | D |
| 18. | C | 38. | D | 58. | B | 78. | A |
| 19. | B | 39. | B | 59. | B | | |
| 20. | B | 40. | A | 60. | C | | |

# Explanations to the Answers for the Sample Test Section 5

Question #1
Answer A is wrong. A defective clutch pulley bearing would make noise when the A/C compressor is disengaged and then go away when the compressor is engaged.
Answer B is wrong. A low refrigerant charge would either cause the compressor to short cycle or to be totally inoperative.
**Answer C is correct.** Internal damage will cause noise only when the clutch is engaged. Noises from an engaged compressor may be caused by an external or internal problem. A compressor noise problem is often difficult to localize. When the compressor is operating, a loose compressor mount or bracket may vibrate, giving the impression that the compressor itself is somehow defective.
Answer D is wrong. An A/C belt that is too tight could cause early failure of the clutch pulley bearing. If this happened, the noise would be present with the compressor disengaged and go away when the compressor is engaged.

Question #2
**Answer A is correct.** Both pressure readings are low; this problem is probably caused by a low refrigerant charge, which is answer A. There are many other possible causes for low readings, but none of the other answer choices would cause them.
Answer B is wrong. These pressures are very low and not in the normal range.
Answer C is wrong. A bad compressor typically causes low high-side pressures and high low-side pressures.
Answer D is wrong. A high-side restriction would cause elevated high-side pressures. The pressures described in the question are both lower than normal.

Question #3
Answer A is wrong. An expansion valve that is stuck open will cause the low-side pressures to be higher than normal and the high-side pressures to be lower than normal.
**Answer B is correct.** The question is looking for the LEAST-Likely cause for low high-side and high low side pressure readings. A restriction on the high side of the system will cause high high-side pressures.
Answer C is wrong. A compressor with faulty reed valves or a bad head gasket will cause high-side pressures to be low and low-side pressures to be high.
Answer D is wrong. A stuck closed POA or STV valve could cause the high-side pressures to be low and the low-side pressures to be high.

Question #4
Answer A is wrong. Mineral oil is used only in R12 systems.
Answer B is wrong. CCOT stands for cycling clutch orifice tube and is not a type of oil.
Answer C is wrong. Ester oil is the type of oil that comes with some retrofit kits.
**Answer D is correct.** The refrigerant oil required by the system depends on a number of things, but it is primarily dictated by the refrigerant used in the system. R-12 systems use a mineral oil. Mineral oil mixes well with R-12 without breaking down. Mineral oil, however, cannot be used with R-134A. R-134A systems require synthetic oil, polyalkylene glycol (PAG). There are a number of different viscosities of PAG oil. Always use the one recommended by the vehicle manufacturer or compressor manufacturer. Failure to use the correct oil will cause damage to the compressor. Many universal retrofit kits include ester oil as a universal oil. Most manufacturers recommend only the proper PAG oil be used in their systems.

Question #5
**Answer A is correct.** Only Technician A is correct. A temperature change in an AC system is caused by a drop in pressure or a change in state of the refrigerant. The receiver/drier should not cause a change in pressure.
Answer B is wrong. The refrigerant in the evaporator absorbs heat it changes state. The outlet of an evaporator is usually around 10° F colder than the inlet.
Answer C is wrong. Only Technician A is correct.
Answer D is wrong Only Technician A is correct.

Question #6
Answer A is wrong. The job of the compressor is to increase the pressure and temperature of the refrigerant vapor.
**Answer B is correct.** Moisture is never needed by the system.
Answer C is wrong. The expansion valve meters refrigerant into the evaporator.
Answer D is wrong. Warm air passing by the evaporator core causes the refrigerant to absorb enough heat to change from a liquid into a vapor.

Question #7
Answer A is wrong. Technician B is also correct.
Answer B is wrong. Technician A is also correct.
**Answer C is correct.** Both technicians are correct. When replacing refrigerant oil, it is important to use the specific type and quantity of oil recommended by the compressor manufacturer. If there is a surplus of oil in the system, too much oil circulates with the refrigerant, causing the cooling capacity of the system to be reduced. Too little oil results in poor lubrication of the compressor. Efficient operation of the air conditioning system greatly depends on the correct amount of refrigerant in the system. A low charge results in inadequate cooling under high heat loads, due to a lack of reserve refrigerant, and can cause the clutch cycling switch to cycle faster than normal. An overcharge can cause inadequate cooling because of a high liquid refrigerant level in the condenser. Refrigerant controls will not operate properly and compressor damage can result. In general, an overcharge of refrigerant will cause higher-than-normal gauge readings and noisy compressor operation.
Answer D is wrong. Both technicians are correct.

Question #8
Answer A is wrong. Technician B is also correct.
Answer B is wrong. Technician A is also correct.
**Answer C is correct.** Both technicians are correct. The problem is that the A/C system seems to work fine in the early morning and evenings, but does not during the heat of the day. During a performance test, the low-side gauge reads normal and then drops dramatically. Both technicians are right. Moisture in the system can be released from the desiccant in the receiver/drier by an increase in temperature on the high side. When the ambient temperature is lower the desiccant may be able to hold the moisture out of the refrigerant. Internal freezing of the expansion valve could produce external icing on the valve.
Answer D is wrong. Both technicians are correct.

Question #9
**Answer A is correct.** The loss of pressure in the cooling system without an evident external leak is probably caused by internal leakage problems. Things such as a leaking transmission cooler, head gasket, or cracked cylinder head could result in coolant loss that is not visually verified. Therefore, all of the answers are valid, except A. A leaking heater core usually causes coolant to leak onto the passenger front floor mat or out of the case drain and is quite apparent. There is also the possibility of a sweet smell when the heater is in operation.
Answer B wrong. Answer B is a true statement but is not the correct answer because this is an EXCEPT question. A leaking internal transmission cooler could cause the pressure tester to decrease without any external signs of leaks. The technician should check the transmission fluid to see if it is contaminated with coolant.
Answer C is wrong. Answer C is a true statement but is not the correct answer because this is an EXCEPT question. A leaking head gasket could cause the pressure tester to decrease without any external signs of leaks. The technician should check the engine oil or the spark plugs for signs of coolant contamination.
Answer D is wrong. Answer D is a true statement but is not the correct answer because this is an EXCEPT question. The technician should check the engine oil or the spark plugs for signs of coolant contamination.

Question #10
**Answer A is correct.** This question gives you the problem and asks what the symptoms of the problem would be. To answer this you need to know the purpose of the vacuum valve in the radiator cap. This valve is designed to reduce the vacuum in the cooling system caused by contraction of the coolant as the hot system cools. A sticking vacuum valve in the radiator cap may allow for a vacuum in the system and collapse the upper radiator hose after the engine is shut off. A malfunctioning vacuum valve would not cause overheating by itself. A loss of coolant could cause overheating.
Answer B is wrong. A vacuum valve that is stuck closed would not cause excessive system pressure because it is only operational when the system cools down and the coolant contracts.
Answer C is wrong. A stuck closed vacuum valve would never cause overheating because it is in the closed position when the engine is under positive pressure.
Answer D is wrong. A stuck closed vacuum valve would not cause overheating during extended idle times because it is supposed to be closed when the system has positive pressure. The only time it opens is when the coolant contracts in the engine and creates a vacuum.

Question #11
Answer A is wrong. The recovery process removes all of the refrigerant but very little oil from an AC system.
Answer B is wrong. The system is ready to be repaired once the recovery process has taken place.
**Answer C is correct.** Recovery is the process of removing refrigerant from the system. The process is identified as complete when a gauge reading of 0 psi or lower can be maintained for 5 minutes. Should the system pressure rise above 0 it indicates the presence of refrigerant and the recovery process should be continued. While some oil may be removed in the recovery process, not all will be. After recovery, the required service should be performed, the oil level replenished to specifications, and a minimum 30-minute vacuum performed. If the system is suspected of being contaminated with moisture, a long (4 hour) vacuum may be required.
Answer D is wrong. Humidity is removed during the evacuation process by lowering the system into a vacuum and allowing the moisture to vaporize and be pulled out of the system.

Question #12
Answer A is wrong. A thermostatic expansion valve (TXV) system would have a receiver/drier.
**Answer B is correct.** A fixed orifice tube system has an accumulator in the suction line.
Answer C is wrong. A receiver/drier is similar to an accumulator but it is found on the high side of a TXV system.
Answer D is wrong. A suction throttling valve (STV) is used on some older TXV systems.

Question #13
Answer A is wrong. The pressure relief valve mounted in either the compressor body or made into the drier. Either location is going to require recovery of the refrigerant for replacement of the valve.
Answer B is wrong. During a retrofit procedure, the pressure relief valve is usually replaced by a high-pressure cut-out switch.
Answer C is wrong. Neither technician is correct.
**Answer D is correct.** Neither technician is correct. A high-pressure relief valve is incorporated into many R-12 air conditioning systems. This valve may be installed on the receiver/drier, compressor, or elsewhere in the high side of the system. It is a high-pressure protection device that opens (normally at 440 psi) to bleed off excessive pressure that might occur in the system. This is necessary to keep system pressures from reaching a point that may cause compressor lockup or other component damage. Technician A is not right. When replacing a high-pressure relief valve, the system must be depressurized before removing the valve. Technician B is not right. It is recommended that you replace the high-pressure relief valve with a high-pressure cutout switch when retrofitting an R-12 system to R-134a. R-134a high-side pressures are normally higher than R-12 systems' and the relief valve may vent refrigerant into the atmosphere.

Question #14
Answer A is wrong. Purging is a process that implies venting refrigerant into the atmosphere and is illegal.
**Answer B is correct.** Only Technician B is correct. Recommended practice within the industry is to always evaluate the system with a refrigerant identifier. This prevents cross contamination of recycled refrigerant and even identifies air in the system that would limit performance.
Answer C is wrong. Only Technician B is correct.
Answer D is wrong. Only Technician B is correct.

Question #15
Answer A is wrong. Technician B is also correct.
Answer B is wrong. Technician A is also correct.
**Answer C is correct.** Both technicians are correct. For the A/C system to be efficient, the evaporator pressure must be maintained to a precise range so that the evaporator temperature will be correct. Some manufacturers utilize a variable displacement compressor to control refrigerant pressure on the high side and evaporator (suction) pressure on the low side. A thermostatic expansion valve (TXV) has been used on many manufacturers' A/C systems to control refrigerant flow and the resulting evaporator pressure. It is a feedback system that regulates flow by use of a temperature-sensing bulb that measures evaporator outlet temperature.
Answer D is wrong. Both technicians are correct.

Question #16
Answer A is wrong. Answer A is a true statement but is not the correct answer because this is an EXCEPT question. It is advisable to pour a few drops of refrigerant oil on top of the orifice tube to lubricate the o-rings.
Answer B is wrong. Answer B is a true statement but is not the correct answer because this is an EXCEPT question. The T-handle should be rotated to engage the notch of the tool in the orifice tube tangs.
**Answer C is correct.** When using the tool to remove an orifice tube, pour a small amount of refrigerant oil on top of the tube and engage the notch in the tool in the orifice tube. Hold the T-handle and rotate the outer sleeve to remove the orifice tube. Turning the T-handle could break off the tabs on the orifice tube and make removal more difficult. Special tools and replacement parts are available to deal with broken orifice tubes.
Answer D is wrong. Answer D is a true statement but is not the correct answer because this is an EXCEPT question. The outer sleeve should be rotated while holding the T-handle during the removal of the orifice tube.

Question #17
Answer A is wrong. Technician B is also correct.
Answer B is wrong. Technician A is also correct.
**Answer C is correct.** Both technicians are correct. During a visual inspection of the system, the components are not only carefully looked at but certain parts are touched while the system is running. This technique can lead you toward a quick diagnosis of the system. The suction line to the compressor should be cool to the touch from the evaporator to the compressor. If it is covered with thick frost, this might indicate that the expansion valve is flooding the evaporator. Technician B is also correct. The high side of the system should never have any cold spots or frost on any components. A formation of frost on the outside of a line or component means there is a restriction to the flow of refrigerant located at the beginning of the frosted point.
Answer D is wrong. Both technicians are correct.

Question #18
Answer A is wrong. Technician B is also correct.
Answer B is wrong. Technician A is also correct.
**Answer C is correct.** Both technicians are correct. A defective radiator cap or restricted radiator tubes may cause engine overheating. When the engine overheats, excessive cooling system temperatures and pressures result. This forces the excessive coolant flow into the recovery container causing the level to rise. As the system cools, it creates a vacuum and draws the coolant from the recovery container back into the radiator.
Answer D is a wrong. Both technicians are correct.

Question #19
Answer A is wrong. An overcharged AC system would elevate systems pressures and reduce cooling performance but it would not cause frosting at the condenser outlet.
**Answer B is correct.** Only Technician B is correct. This vehicle is equipped with a TXV system and has excessive high pressure. Technician A suggests the system may be overcharged. If the system were overcharged, the head pressure may be excessive but there would be no frosting at the condenser outlet. Technician B understands that frost will build up at a point of restriction. Frosting near the condenser outlet is an indication that there is a restriction in the condenser.
Answer C is wrong. Only Technician B is correct.
Answer D is wrong. Only Technician B is correct.

Question #20
Answer A is wrong. Most cabin air filters do not filter the re-circulated air.
**Answer B is correct.** A cabin air filter removes particles from the outside air entering the AC system. Re-circulated air should already be cleaned by the cabin air filter as it entered the car. All sources and heated air only are ruled out by the filter being in the outside source, which supplies air for ventilation, heating, and AC operation.
Answer C is wrong. Most cabin air filters do not filter all air sources. They typically just filter the fresh air.
Answer D is wrong. Cabin air filters protect more than just the air passing by the heater core.

Question #21
Answer A is wrong. Answer A is a true statement but is not the correct answer because this is an EXCEPT question.
Answer B is wrong. Answer B is a true statement but is not the correct answer because this is an EXCEPT question.
**Answer C is correct** .All technicians should know the differences between R-12 and R-134A systems before working on these systems. Since R-134A is not interchangeable with R-12, separate sets of hoses, gauges, and other equipment are required to service vehicles. All equipment used to service R-134A and R-12 systems must meet SAE standard J1991. The service hoses on the manifold gauge set must have manual or automatic back-flow valves at the service port connector ends. This prevents the refrigerant from being released into the atmosphere during connection and disconnection. Manifold gauge sets for R-134A can be identified by one or all of the following: Labeled *FOR USE WITH R-134A, Labeled HFC-134 or R-134A,* and/or have a light blue color on the face of the gauges. For identification purposes, R-134A service hoses must have a black stripe along their length and be clearly labeled *SAE J2196/R-134A.* The low-pressure hose is blue with a black stripe. The high-pressure hose is red with black stripe and the center service hose is yellow with a black stripe. Service hoses for one type of refrigerant will not easily connect into the wrong system, as the fittings for an R-134A system are different than those used in an R-12 system. Answer C is the EXCEPT answer because there are not significant operational pressure differences between R-12 and R-132a systems.
Answer D is wrong. Answer D is a true statement but is not the correct answer because this is an EXCEPT question.

Question #22
Answer A is wrong. Technician B is also correct.
Answer B is wrong. Technician A is also correct.
**Answer C is correct.** Both technicians are correct. The presence of oil around any A/C fitting or component usually indicates a leak. However if oil is present around the high-pressure relief valve, this indicates the valve is opening. A high-pressure relief valve is used to keep system pressures from reaching a point that may cause compressor lockup or other component damage because of excessive high pressures. When system pressures exceed a predetermined point, the pressure relief valve opens, allowing excessive system pressures to be reduced. High-system pressures may be caused by restricted condenser air passages or an overcharged refrigerant system.
Answer D is wrong. Both technicians are correct.

Question #23
Answer A is wrong. There would not have been voltage at terminal C if the thermal fuse were open.
**Answer B is correct.** The choice of open clutch coil is best because voltage is available at terminal C, which feeds the clutch, and the clutch is inoperative. If the thermal fuse had been open, there would be no voltage at terminal C. The normally open superheat switch provides a ground path for the heater element when the specified compressor temperature is reached, causing the thermal fuse to blow. A defective heater would cause the thermal fuse to fail to open when the superheat switch closed.
Answer C is wrong. A defective superheat switch would cause the thermal fuse to fail to open in case of high compressor temperatures.
Answer D is wrong. A defective thermal fuse heater would cause the thermal fuse to fail to open in case of high compressor temperatures.

Question #24
Answer A is wrong. Technician B is also correct.
Answer B is wrong. Technician A is also correct.
**Answer C is correct.** Both technicians are correct. A faulty cycling switch that is stuck closed would cause continuous compressor operation and could result in the conditions described. A clogged orifice tube would not allow normal flow of refrigerant into the evaporator and the low-side gauge reading after the evaporator would be lower than normal, possibly even a vacuum.
Answer D is wrong. Both technicians are correct.

Question #25
Answer A is wrong. Answer A is a true statement but is not the correct answer because this is an EXCEPT question. A stuck-open thermostat would definitely cause poor heater performance.
Answer B is wrong. Answer B is a true statement but is not the correct answer because this is an EXCEPT question. A stuck-open thermostat would definitely cause cooler engine temperatures.
Answer C is wrong. Answer C is a true statement but is not the correct answer because this is an EXCEPT question. A stuck-open thermostat would cause increased fuel consumption because the engine would take longer to build engine temperature and could not sustain normal operating temperature; especially in colder climates.
**Answer D is correct.** A stuck-open thermostat would not cause overheating. An open thermostat would be the normal position for a warmed-up engine. The engine thermostat speeds engine warm-up by restricting coolant flow until a predetermined temperature is reached. Poor heater operation and excess fuel consumption would be caused by slow warm-up. The thermostat also serves to regulate coolant flow, which keeps the engine at an acceptable and efficient operating temperature.

Question #26
Answer A is wrong. Technician B is also correct.
Answer B is wrong. Technician A is also correct.
**Answer C is correct.** Both technicians are correct. Technician A is correct because a V belt that is not wide enough will run on the bottom of the pulley providing very little belt friction regardless of the correct tension. V belts get their friction by wedging themselves into the V groove. Technician B is also correct because an overcharged condition can cause the high-side pressure to be so high as to resist the start-up of the compressor. In the long run, the clutch or belt will usually fail if the system continues to operate with excessive high-side pressures.
Answer D is wrong. Both technicians are correct.

Question #27
Answer A is wrong. Technician B is also correct.
Answer B is wrong. Technician A is also correct.
**Answer C is correct.** Both technicians are correct. A replacement compressor should have the same type of clutch and pulley as the old compressor. The mounting brackets and other fasteners on a replacement compressor should also be identical to those on the old compressor.
Answer D is wrong. Both technicians are correct.

Question #28
Answer A is wrong. Modern compressors no longer have a dipstick to allow the compressor oil to be checked.
**Answer B is correct.** Only Technician B is correct. All modern compressors require the oil be drained from the compressor and measured in a graduated container. If the system had a leak, the measured amount of oil may be lower than specifications. Always refer to the guidelines given by the manufacturer before adding oil.
Answer C is wrong. Only Technician B is correct.
Answer D is wrong. Only Technician B is correct.

Question #29
Answer A is wrong. A defective TXV capillary tube would affect the cooling capacity at all times.
Answer B is wrong. A refrigerant overcharge would cause reduced performance at all times.
Answer C is wrong. A restricted condenser passage would cause reduced performance at all times and would have frost in the area right past the restriction.
**Answer D is correct.** A system that stops cooling after a period of operation could be caused by several problems. Of the choices given, a TXV capillary tube would not be affected solely by period of operation. If it were defective it would affect cooling at all times. A refrigerant overcharge and condenser restriction is going to affect system operation from the beginning of the system's operation. With the conditions given, the possibility that moisture in the refrigerant system would create an ice blockage at the orifice tube or TXV is the most probable choice. After the system is shut down the ice would melt, allowing the system to resume normal operation until the ice blockage occurs again. The desiccant in the receiver/drier or accumulator is intended to absorb and hold moisture to prevent this problem. Saturation of the desiccant allows moisture to circulate with the refrigerant.

Question #30
Answer A is wrong. Technician B is also correct.
Answer B is wrong. Technician A is also correct.
**Answer C is correct.** Both technicians are correct. The problem is that the A/C system seems to work fine in the early morning and evenings, but does not during the heat of the day. During a performance test, the low-side gauge reads normal and then drops dramatically. Both technicians are right. Moisture in the system can be released from the desiccant in the receiver/drier by an increase in temperature on the high side. When the ambient temperature is lower the desiccant may be able to hold the moisture out of the refrigerant. Internal freezing of the expansion valve could produce external icing on the valve.
Answer D is wrong. Both technicians are correct.

Question #31
**Answer A is correct.** Since the condenser is a heat exchanger that causes the refrigerant to condense back to liquid, it will not be very effective when the fins get restricted with road debris. This causes system efficiency (cooling) to drop off. Should the high-side pressure become high enough, a relief valve on some R-12 systems will vent refrigerant to control the pressure. Newer R-134a systems use a high-pressure cutout switch to prevent the high-side pressures from getting too high.
Answer B is wrong. Freezing of the evaporator is caused be a problem with the de-icing components in the A/C system.
Answer C is wrong. Poor airflow could be the result of a blower problem or a restricted cabin air filter.
Answer D is wrong. The loss of refrigerant is typically caused by a leak in the system..

Question #32

Answer A is wrong. Restricted airflow in the condenser would cause very high system pressures and a lack of performance but it would not cause frost at any point on the condenser.

Answer B is wrong. A refrigerant leak in the condenser would cause an oily spot at the location of the leak.

**Answer C is correct.** When frost is forming on one of the condenser tubes near the bottom of the condenser, the refrigerant passage is restricted at that location in the condenser. Frosting is a clear indication that there is a restriction in the condenser. Restricted airflow passages in the condenser would not cause frosting; it would limit the efficiency of the system because the refrigerant would not be cooled by the air flowing through the condenser. A leak would not cause frosting either; oil may be seen at the point of leakage. If the orifice tube were restricted, there would be frost around it.

Answer D is wrong. A restricted orifice tube would cause low system pressures and lower performance of the A/C system but it would not cause frosting on the condenser.

Question #33

Answer A is wrong. The valve is normally closed which means that it gets no vacuum when the HVAC controls are set for full cold.

Answer B is wrong. Debris in the heater core inlet would cause reduced heater output but would not affect A/C operation.

**Answer C is correct.** A stuck-open heater control valve would allow the heater core to add heat to the cooled air reducing the effectiveness of the A/C system. With a normally closed heater control valve, vacuum would be applied to the valve to cause it to open and add heated air. Vacuum would not be applied to the valve in A/C mode so a leak would not cause this problem. A stuck-closed valve would restrict coolant flow to the heater core and the complaint would be no heat, not extra heat.

Answer D is wrong. The heater control valve is normally closed which means that it gets no vacuum when the HVAC controls are set for full cold. A stuck closed valve would reduce heater performance but not affect A/C operation.

Question #34

**Answer A is correct.** Only Technician A is correct. When you examine the vacuum diagram, in the heat position the vacuum switch would be in the #5 position and vacuum would be supplied to the top of the vacuum motor causing the door to move to the heater outlet position.

Answer B is wrong. Vacuum is only supplied to the panel door valve actuator in the Max, Normal, and Vent positions.

Answer C is wrong. Only Technician A is correct.

Answer D is wrong. Only Technician A is correct.

Question #35

**Answer A is correct.** Only Technician A is correct. An inoperative Bowden cable-controlled heater control valve moves freely, but the valve doesn't respond to the change. Technician A is the correct choice in saying the cable housing clamp may be loose at the control head (panel) end.

Answer B is wrong. If the cable were rusted in the housing, the control would not move freely.

Answer C is wrong. Only Technician A is correct.

Answer D is wrong. Only Technician A is correct.

Question #36

Answer A is wrong. If the pressure lowered when the remote bulb is cooled, then the TXV is working correctly.

**Answer B is correct.** Only Technician B is correct. In this question, the low-side pressure is higher than normal. When the expansion valve's remote bulb or capillary tube is cooled, the pressure decreases indicating that the valve is reducing refrigerant flow into the evaporator. If the expansion valve were faulty, it probably would not respond to the change in temperature. If the expansion valve were defective, cooling the remote bulb would make little or no difference in the low-side pressure. If there was no difference in pressure, one may assume that the valve was stuck in an open position and not throttling down to properly meter refrigerant into the evaporator. If there is no indication that the expansion valve is defective, the remote bulb is Most-Likely not making proper contact with the evaporator tail pipe and not sensing the correct temperature.

Answer C is wrong. Only Technician B is correct.

Answer D is wrong. Only Technician B is correct.

Question #37

Answer A is wrong. A restricted orifice tube would cause a lower- (not higher) than-normal low-side pressure.
**Answer B is correct.** Only Technician B is correct in saying a restricted orifice tube may cause frosting of the tube. The restriction will cause a greater than normal pressure drop that can lead to the refrigerant vaporizing at that point. Vaporization of the refrigerant will produce cold enough temperatures to cause frosting.
Answer C is wrong. Only Technician B is correct.

Question #38

Answer A is wrong. A technician should never charge into the high side with the engine running. Doing this could result in the high side pressure being pumped into the charging cylinder and exploding.
Answer B is wrong. Inverting a charge cylinder results in liquid refrigerant being sent to the manifold set.
Answer C is wrong. Neither Technician A nor B is correct.
Answer D is correct. Neither technician is correct. An A/C system should only be charged through the high side with the system off. Attempting to charge through the high side with the system in operation can be very dangerous because the charging cylinder may be pressurized to the point of rupture. Inverting the refrigerant container disperses liquid refrigerant, which could cause damage to the compressor. The vehicle should always be charged from the low-pressure side if the engine is on. Heating of the refrigerant within acceptable limits can speed up the process of low-side vapor charging.

Question #39

Answer A is wrong. A plugged heater case drain would not cause an oily film on the windshield. An oily film could only be cause by a refrigerant leak or by a heater core leak. A plugged case drain could cause water to leak out onto the carpet or possibly a musty smell coming from the vents.
Answer B is correct. Only Technician B is correct. An oil film on the windshield could be caused by a refrigeration leak in the evaporator core that may allow some refrigerant oil to escape into the evaporator case. Keep in mind, the presence of oil on an A/C component or line is a good indication of a leak. AC/heater blower operation in the defrost mode could carry the oil through the ducts to the windshield.
Answer C is wrong. Only Technician B is correct.
Answer D is wrong. Only Technician B is correct.

Question #40

**Answer A is correct.** The tool shown in the figure is a special tool used to release the spring lock couplings on refrigerant lines. These line connections are sealed with internal O-rings and retained with spring lock couplings. These fittings speed assembly during manufacture because they are snapped together by hand but require this tool to disconnect them.
Answer B is wrong. No tool is needed to connect spring lock couplings.
Answer C is wrong. The tool is only needed to disconnect spring lock couplings.
Answer D is wrong. Male barb type fittings are crimped by a hydraulic powered crimping machine.

Question #41

**Answer A is correct.** In the past, the evaporator temperature regulator (ETR) and evaporator pressure regulator (EPR) valves were popular methods of evaporator temperature control, but these have not been in use for several years. Technician B is not right. Suction throttling valve (STV) is the term used for all devices that were used in the suction line from the evaporator to the compressor to maintain evaporator pressure to maintain a constant evaporator temperature. All of these valves were used on older cars and were replaced with orifice tube, cycling clutch, and variable displacement compressors on newer systems.
Answer B is wrong. STV systems were used on older A/C systems to help control evaporator temperature.
Answer C is wrong. Only Technician A is correct.
Answer D is wrong. Only Technician A is correct.

Question #42

**Answer A is correct.** A flame-type (halide) leak detector is effective at finding leaks on R-12 systems but is ineffective on R-134a systems. A halogen or electronic leak detector can identify most leaks when used properly. A special fluorescent dye added to the refrigerant oil of the system will leak out along with the oil. An ultraviolet light can be used to see the point of the leak. The system may have to be operated for several days if the leak is small. Some leaks are large enough that a sound like a tire leak can be heard. Special listening tools can enhance the possibility of finding an audible leak.

Answer B is wrong. Halogen leak detectors are useful in finding leaks in R134a systems when used correctly.

Answer C is wrong. The dye method of leak detection is very effective on R12 or R134a systems.

Answer D is wrong. Some leaks are large enough to hear an audible hiss from the leak location.

Question #43

Answer A is wrong. Technician B is also correct.

Answer B is wrong. Technician A is also correct.

**Answer C is correct.** Both technicians are correct. Technician A is correct because compressors are equipped with an electromagnetic clutch as part of the compressor pulley assembly. The clutch is designed to engage the pulley to the compressor shaft when the clutch coil is energized. Nearly all clutch assemblies have a clearance specification for the distance between the clutch and the pressure plate. This clearance is measured with a feeler gauge as indicated in the diagram. If the clearance is too great, the clutch may slip intermittently and cause a scraping or squealing noise. Technician B is also correct because the scraping noise is the sound made by a slipping compressor clutch.

Answer D is wrong. Both technicians are correct.

Question #44

Answer A is a good choice but not the correct answer because both technicians are correct.

Answer B is a good choice but not the correct answer because both technicians are correct.

**Answer C is correct.** Schrader-type service valves provide a simple connection similar to a tire valve. When manifold gauge set hoses are connected to the Schrader type, a pin in the hose depresses the Schrader valve so the valve will open, allowing refrigerant to flow into or out of the sealed A/C system. Another type valve, the stem type, has a manual valve that can be opened once the manifold gauge set is installed and closed before removing the manifold gauge set. This allows the compressor to be isolated from the system for service or oil level checks.

Answer D is wrong because both technicians are correct.

Question #45

Answer A is wrong. Technician B is also correct. a good choice but not the correct answer because both technicians are correct.

Answer B is wrong. Technician A is also correct. a good choice but not the correct answer because both technicians are correct.

**Answer C is correct.** Both technicians are correct. A noise from the area of the compressor can be the result of loose or broken fasteners. A compressor that can be moved on its mount is a sign that there is something loose or broken in the compressor mounts.. All fasteners and brackets should be checked.

Answer D is wrong. Both technicians are correct.

Question #46

Answer A is wrong. Technician B is also correct.

Answer B is wrong. Technician B is also correct.

**Answer C is correct.** Both technicians are correct. The desiccant used in receiver/driers and accumulators does draw moisture and it can become contaminated in a few minutes if it is exposed to the atmosphere. When replacing the desiccant, keep the new bag sealed. Once opened, put it under a vacuum quickly. The accumulator is connected into the low side, at the outlet of the evaporator. The accumulator also contains a desiccant and is designed to store excess refrigerant and to filter and dry the refrigerant. If liquid refrigerant flows out of the evaporator, it will be collected by and stored in the accumulator. The main purpose of an accumulator is to prevent liquid from entering the compressor.

Answer D is wrong. Both technicians are correct.

Question #47

Answer A is wrong. Technician B is also correct.

Answer B is wrong. Technician A is also correct.

**Answer C is correct.** Both technicians are correct. A sticky film on the inside of the windshield may be caused by a coolant leak in the heater core. Under this condition, the coolant level in the radiator should be checked. The heater core should be repaired or replaced. A leaking heater core is also often the cause of a windshield fogging problem. Windshield fogging is the result of hot and humid air that condenses on the cooler glass.

Answer D is wrong. Both technicians are correct.

Question #48

**Answer A is correct.** As pressure increases in the cooling system, the boiling point of the coolant is also increased. Similarly, when more antifreeze is added to the coolant, the boiling point is increased. Most ethylene glycol antifreeze contains rust and corrosion inhibitors. Coolant is considered a hazardous waste and must be recovered. Any coolant removed from an automotive cooling system may contain harmful engine residues, such as benzene and lead. If used coolant is to be disposed of, it must be done according to local regulations. Ideally, however, the coolant may be recycled by filtering out most of the contaminants. Some recycle machines use vacuum distillation or ion exchange technology for glycol-based coolants. After recycling, the coolant must be supplemented with an additive before being reused.

Answer B is wrong. Adding more antifreeze to the coolant mixture does increase the boiling point to a limited extent.

Answer C is wrong. Quality ethylene glycol antifreeze does contain antirust and corrosion inhibitors. These inhibitors help the coolant maintain positive qualities for a longer time.

Answer D is wrong. Answer D is a true statement but is not the correct answer because this is an EXCEPT question.

Question #49

Answer A is wrong. The TXV should never freeze up during normal operation. Ice buildup on the TXV usually indicates that the TXV is stuck closed.

Answer B is wrong. A restriction in the receiver/drier would produce an unwanted temperature drop at that location.

Answer C is wrong. Neither technician is correct.

**Answer D is correct.** Neither technician is correct. Technician A says the TXV is the coldest part of the system and heavy ice buildup on it is normal. This is wrong. The thermostatic expansion valve regulates the flow of refrigerant to cause a pressure and temperature drop in the refrigerant. Heavy ice buildup at the TXV is not normal and usually indicates that the refrigerant is changing state at that point rather than in the evaporator. Technician B is also wrong. A restriction in the receiver/drier would produce an abnormal drop in temperature and possibly ice at that spot. A clear sight glass on the receiver/drier does not indicate there is no refrigerant flowing. It indicates that the system is either fully charged or empty. Bubbles in the sight glass do not always indicate a low refrigerant charge.

Question #50

Answer A is wrong. A blown cooling fan fuse would prevent the cooling fan motors from operating even with the A/C on.

**Answer B is correct.** A defective engine coolant temperature sensor that indicates low coolant temperature may cause the power train control module (PCM) not to ground the fan relay windings and thus prevent the operation of the cooling fans except when the A/C pressure switch closes. The problem in this question is inoperative cooling fans when the engine is warm or hot, but the fans work when the A/C is turned on. To diagnose this problem, look at the schematic and determine the parts of the circuit that control the fans based on engine temperature. That is where the problem will be. The circuit tied to the A/C system appears to be working fine.

Answer C is wrong. A defective high-speed coolant fan relay would only prevent high-speed fan operation.

Answer D is wrong. A defective A/C pressure fan switch would keep the fan from operating when the A/C is turned on. The question states that both fans work when the A/C is on.

Question #51

**Answer A is correct.** To answer this question, look at the figure and identify the parts shown. Then identify what is being done. You will notice the linkage to the air mix or blend door is being adjusted. This adjustment is critical to the proper operation of the system. The blend system mixes heated air with cooled air to achieve the desired air temperature. When the A/C system is set to max, only cooled air should leave the air ducts. When full heat is selected, only heated air should be available.
Answer B is wrong. There is no coolant control valve in the picture.
Answer C is wrong. There is no ventilation door control rod in the picture.
Answer D is wrong. There is no defroster door control rod in the picture.

Question #52

**Answer A is correct.** Only Technician A is correct. The function of a blower resistor is to slow down the blower to various speeds by way of a voltage drop.
Answer B is wrong. Only Technician A is correct. In the high position, the resistor is bypassed by a relay on many systems and full voltage is supplied to the motor. There are multiple resistance taps on the resistor block that make up each speed. Remember the more resistance, the greater the voltage drop and the lower the blower speed.
Answer C is wrong. Only Technician A is correct.
Answer D is wrong. Only Technician A is correct.

Question #53

**Answer A is correct.** Only Technician A is correct. To answer this question you need to think about what could cause the blower motor not to work. To do this, think about the blower circuit and only the blower circuit. To function properly, the blower motor requires a complete electrical circuit. It is generally connected to the negative side of the battery via the frame of the vehicle and to the positive side of the battery via insulated wires, speed resistors, the switch, the fuse or circuit breaker, and the ignition switch. An open circuit in the resistor or resistor terminals at the switch will cause the blower motor to operate at a speed not selected or not at all. An open at the switch terminal of the resistor will not allow the blower motor to operate at the speed selected by the switch. An open circuit at the blower switch ground or at the terminal causes the blower to be completely inoperative. As Technician A said, a loose ground wire will cause an inoperative blower motor.
Answer B is wrong. The A/C compressor clutch diode has no affect on the blower motor. This clamping diode is used to suppress the voltage spike from the A/C clutch coil.
Answer C is wrong. Only Technician A is correct.
Answer D is wrong. Only Technician A is correct.

Question #54

Answer A is wrong. There should be no alternating current in any of the electrical systems on a vehicle. The alternating current in the alternator is rectified to direct current before exiting the alternator.
**Answer B is correct.** Only Technician B is correct. The purpose of the diode in the compressor clutch circuit is to prevent spikes of high voltage from returning to the delicate electronic components of the computer. The clutch coil is an electromagnet with a strong magnetic field when current is applied. This magnetic field is constant as long as power is applied to the coil. When power is removed, the magnetic field collapses and creates high-voltage spikes. These are harmful to the delicate electronic circuits of the computer and must be prevented. The diode placed across the clutch coil provides an electrical path to ground, holding the spikes to a safe level.
Answer C is wrong. Only Technician B is correct.
Answer D is wrong. Only Technician B is correct.

Question #55

**Answer A is correct.** Only Technician A is correct. An ambient temperature switch senses outside air temperature and is designed to prevent compressor clutch engagement when air conditioning is not required.
Answer B is wrong. The ambient temperature switch opens at approximately 37 degrees and remains open until the temperature rises above 37.
Answer C is wrong. Only Technician A is correct.
Answer D is wrong. Only Technician A is correct.

Question #56

Answer A is wrong Technician B is also correct.

Answer B is a wrong. Technician A is also correct.

**Answer C is correct.** Both technicians are correct. During acceleration or when pulling heavy loads the AC compressor is cut off to provide additional engine power. Some vehicles have a power steering cutoff switch to disengage the air conditioning compressor when the power steering requires maximum effort. This prevents the engine from stalling during parking maneuvers Some systems will disengage the compressor during other programmed conditions where the AC operation is of secondary importance, such as engine overheating.

Answer D is wrong. Both technicians are correct.

Question #57

**Answer A is correct.** The cooling fan operates normally on high speed but there is no low-speed fan operation under any operating condition. A look at the wiring diagram shows there are two separate relays and control circuits for low- and high-speed operation. We can assume that everything on the high-speed part of the circuit is fine, so let's analyze the answers. If there is an open circuit at pin 30 of the low speed relay, the low-speed relay would not receive power and the fan low-speed would not work. Answers B and C say a blown cooling fan fuse would cause the cooling fans to be inoperative at all times. Therefore, both of these are wrong. Answer D suggests an open in the pressure switch circuit. If the cooling fan only operates when A/C pressure is high, suspect an open A/C head pressure switch. However, since the fan operates normally at high speed, the lack of low-speed operation is probably not caused by an open in the pressure switch circuit.

Answer B is wrong. A blown 40 amp maxifuse would prevent both fans from operating.

Answer C is wrong. A blown 10 amp fuse would prevent both fans from operating.

Answer D is wrong. An open at the A/C pressure switch would only affect cooling fan operation during A/C operation.

Question #58

Answer A is wrong. Some climate control systems allow calibration of the actuator motors by putting the control head in self-diagnostic mode.

**Answer B is correct.** Here is a summary of computer-controlled A/C system actuator motors. Some of these motors are calibrated automatically in the self-diagnostic mode, whereas other actuator motors must be calibrated manually. These actuators should only require calibration after motor replacement or misadjustment. Answer B is the correct choice because it makes an incorrect statement. Some systems have a self-diagnosis mode. The diagnostic trouble codes do indicate faults in specific components or circuits.

Answer C is wrong. Some climate control systems require manual calibration of the actuator motors.

Answer D is wrong. The only time that the actuator motors should need to be calibrated is after motor replacement or misadjustment.

Question #59

Answer A is wrong. It is always advisable to disconnect the negative battery cable before servicing any electronic component.

**Answer B is correct.** This question contains four statements about control panel service. Your task is to find the incorrect statement. When inspecting, testing, servicing, or replacing heating, ventilating, and A/C control panel assemblies, disconnect the negative battery cable and wait the specified length of time before working on the control panel. This is especially true if you are working on a vehicle with an air bag system. The disconnecting of the battery and the wait helps prevent unwanted air bag deployment. Self-diagnostic tests of the system may indicate a defective A/C control panel. Answer B is not a true statement and therefore is the correct answer for this question. The refrigeration system does not require discharging before A/C panel removal. The control panel is electrical and possibly mechanical but is not connected to the refrigerant system.

Answer C is wrong. Air bags require the technician to wait a specified time after disconnecting the negative battery cable in order for the capacitors to discharge in the airbag ECU.

Answer D is wrong. Computer-controlled climate control systems are usually capable of running diagnostic tests on many components of the air conditioning system, including the control head.

Question #60
Answer A is wrong. Technician B is also correct.
Answer B is wrong. Technician A is also correct.
**Answer C is correct.** A disconnected vacuum supply line would cause all of the vacuum controls to be inoperative. If the manifold vacuum fitting is restricted, the controls will not operate due to lack of a vacuum signal. A single inoperative vacuum motor could be due to a loss of the vacuum signal for that device only. This could be the fault of a defective vacuum switch, a clogged check valve, a disconnected or damaged hose, a clogged restrictor, a restriction in the system, or a defective vacuum motor.
Answer D is wrong. Both technicians are correct.

Question #61
Answer A is wrong. Compressor operation is controlled by an input from the climate control panel.
Answer B is wrong. The heater control valve is opened and closed as a result of control panel operation.
Answer C is wrong. Plenum door position is moved as the control panel settings are changed.
**Answer D is correct.** The control panel assembly controls the compressor, heater valve, and plenum door. The control panel assembly has no control over the engine coolant temperature nor does it control the coolant temperature sensor. The purpose of the control panel is to provide operator input for the air conditioning and heating system. Some control panels have provisions for displaying in-car and outside air temperatures. A microprocessor may be located in the control panel to provide input data to the programmer, based on operator-selected conditions.

Question #62
Answer A is wrong. It is never ok to replace a part before completely diagnosing the circuit.
Answer B is wrong. Disconnecting the battery will cause the loss of stored data which could help in diagnosing the code.
Answer C is wrong. Neither technician is correct.
**Answer D is correct.** Neither technician is correct. When a fault code is obtained representing the temperature blend door, the first step in the diagnostic procedure should be to check the temperature blend door for a sticking condition.

Question #63
Answer A is a wrong. Technician B is also correct.
Answer B is wrong. Technician A is also correct.
**Answer C is correct.** Both technicians are correct. A gurgling noise in the heater core may be caused by a low coolant level in the cooling system or a restricted heater core. A low coolant level will allow too much air in the cooling system. The excessive air in the system will mix with the coolant and create the gurgling noise. A restricted heater core will also cause a gurgling noise from the coolant passing through the restricted area.
Answer D is wrong. Both technicians are correct.

Question #64
Answer A is wrong. A leaking panel actuator would cause a problem all of the time and not just when the vehicle is under heavy load.
**Answer B is correct.** A leaking check valve will not trap the vacuum in the reserve tank when climbing a long hill. This action may cause the temperature blend door to move to the warm air position and the air discharge to switch to the floor ducts. All of the other answers are problems that would be present all of the time and not just when the engine was heavily loaded.
Answer C is wrong. A leaking blend-air door actuator would cause air temperature problems constantly.
Answer D is wrong. A leaking intake manifold gasket would cause low vacuum at all times and would constantly affect mode and temperature door operation.

Question #65
**Answer A is correct.** Only Technician A is correct. When the outside recirculation door is in position A, only outside air is allowed to enter the air conditioner heater case. When the outside recirculation door is in position B, only inside air is allowed to enter the air conditioner heater case.
Answer B is wrong. The air-recirculation door would have no affect on the heater operation. The temperature blend door is what causes the air to be either routed through or around the heater core.
Answer C is wrong. Only Technician A is correct.
Answer D is wrong. Only Technician A is correct.

Question #66
Answer A is wrong. Technician B is also correct.
Answer B is wrong. Technician A is also correct.
**Answer C is correct.** If the in-car temperature sensor is defective, it may be sending a temperature signal that is colder than requested. This would cause the A/C temperature to be reduced to compensate for the cooler reading and thereby raise the in-car temperature. A sticking temperature blend door will cause warm air to be mixed with cool air and result in the in-car temperature being higher than the driver-selected temperature.
Answer D is wrong. Both technicians are correct.

Question #67
**Answer A is correct.** The problem in this question is that the blower motor is inoperative at all speeds. If voltage was present on terminal D2, then the most likely possible answer would be a defective blower motor.
Answer B is wrong. If this fuse was blown, then there would be no voltage available to the power module.
Answer C is wrong. If the power module was defective, then there would be no voltage available on terminal D2.
Answer D is wrong. Circuit D2 is getting a voltage signal from the power module.

Question #68
Answer A is wrong. If the cooling fan was inoperative, the vehicle would overheat but the vented air would be not be heated.
Answer B is wrong. If the coolant level was low, the vehicle would overheat and, like answer A, the vented air from outside the car would not be heated.
Answer C is wrong. If the heater core were clogged, there would be little or no heat.
**Answer D is correct.** A stuck blend door would allow vented air to pass through the heater core and pick up heat.

Question #69

Answer A is wrong. Answer A is a true statement but is not the correct answer because this question was looking for the answer that is NOT TRUE.

**Answer B is correct.** The following guidelines should be followed when converting an older A/C system to R-134A. These guidelines should allow you to provide the customer with a cool vehicle and to meet current legislative mandates. The guidelines are listed in order and reflect the necessary steps for making this conversion.

- Visually inspect all air conditioning and heater system components.
- Use a refrigerant identifier to make sure the system only contains pure R-12.
- Leak check the system for leaks.
- Run a performance test and record the temperature and pressure readings.
- Remove all R-12 from the system with a recycling machine.
- If the system uses a compressor with an oil sump, remove the compressor and drain all oil from it. Measure the amount of oil drained out.
- Remove and inspect the expansion valve or the orifice tube; replace it if necessary. If either is contaminated, flush the condenser.
- Remove the filter/drier or accumulator, drain it, and measure the amount of oil in it.
- Before converting, make all necessary system changes, such as hoses, gaskets, and seals.
- Install R-134A compatible oil into the system; put in the same amount you took out.
- Install a new filter/drier or accumulator with the correct desiccant.
- Permanently install conversion fittings using a thread-locking chemical.
- Install a high-pressure cutoff switch if the system does not have one.
- Install conversion labels on top of or remove the R-12 label. The R-12 label should not be visible.
- Connect R-134A system evacuation equipment and evacuate for at least 30 minutes.
- Recharge the system to approximately 80 percent of the original R-12 charge.
- Run a performance check and compare readings with those taken before the conversion.
- Leak check the system.

Answer C is wrong. Answer C is a true statement but is not the correct answer because this question was looking for the answer that is NOT TRUE.

Answer D is wrong. Answer D is a true statement but is not the correct answer because this question was looking for the answer that is NOT TRUE.

Question #70

**Answer A is correct.** Only Technician A is correct. Restricted evaporator refrigerant passages may cause frosting at the restriction or the evaporator outlet pipe.

Answer B is wrong. Restricted passages in the evaporator would cause lower than normal low-side pressures.

Answer C is wrong. Only Technician A is correct.

Answer D is wrong. Only Technician A is correct.

Question #71

Answer A is wrong. If the pressure cut-off switch was stuck open, then the compressor would be inoperative.

Answer B is wrong. The coil is wired directly to ground either by a case ground on the coil or by a ground wire connected to the coil.

Answer C is wrong. If the low pressure switch had an open wire, then the compressor would be inoperative.

**Answer D is correct.** If the positive wire to the compressor clutch coil is shorted to a 12-volt source, the compressor clutch will not disengage. This is simply because there is no control of the power to the clutch.

Question #72
Answer A is wrong. The question states that the hose is hot between the heater valve and the heater core. This indicates that the heater valve is supplying the heater core with hot coolant.
**Answer B is correct.** Only Technician B is correct. The indication is that hot coolant is passing the control valve but not exiting the heater core. Therefore, the heater core may be clogged and should be cleaned or replaced.
Answer C is wrong. Only Technician B is correct.
Answer D is wrong. Only Technician B is correct.

Question #73
Answer A is wrong. The equipment label must indicate UL approval.
Answer B is wrong. The equipment label must indicate SAE J1991 approval.
**Answer C is correct.** The properly matched recovery container must be used with recovery/ recycling equipment. The refrigerant container specified by the recovery/recycling equipment manufacturer must be used to ensure that the container has proper capacity, weight, sensor connections, and valving.
Answer D is wrong. Refrigerant oils must not be mixed during the recovery/recycling process.

Question #74
**Answer A is correct.** If, after the recovery process, the low-side gauge rises above 0 psi, there is still some refrigerant remaining in the system. Any remaining refrigerant would evaporate and cause the system pressure to rise above atmosphere. This would show as above 0 psi because atmosphere would show on the gauge as 0 psi. The system should be evacuated until there is a vacuum left in the system.
Answer B is wrong. Oil left in the system would not cause a pressure above zero.
Answer C is wrong. A leak in the system would not cause the pressure to rise above zero psi after the recovery process.
Answer D is wrong. Excessive moisture in the system would not cause the pressure to rise above zero psi after the recovery process.

Question #75
Answer A is wrong. During the recovery process the machine may be able to measure the amount of refrigerant recovered, but it is not to correct the amount of charge in the system. Knowing how much refrigerant is recovered is an essential function in diagnosing some system problems.
**Answer B is correct.** Only Technician B is correct. When the recovery tank reaches 80% capacity the machine should shut down until the recovery cylinder is replaced. If the cylinder is filled beyond 80% it could burst when the refrigerant temperature increases. Some machines will also shut down when an internal container reaches a limit of recovered oil.
Answer C is wrong. Only Technician B is correct.
Answer D is wrong. Only Technician B is correct.

Question #76
Answer A is wrong. Compressor operation is controlled by an input from the climate control panel.
Answer B is wrong. The heater control valve is opened and closed as a result of control panel operation.
Answer C is wrong. Plenum door position is moved as the control panel settings are changed.
**Answer D is correct.** Refrigerant storage containers must not be filled to greater than 80 percent of their gross weight rating. If this level passed, the refrigerant could expand when heated to greater than 100 percent volume and cause the container to rupture or leak.

Question #77
Answer A is wrong. The refrigerant needs to be stored for twelve hours before testing.
Answer B is wrong. The container should not be stored near any heat source. The shop window will allow sunlight to heat the container from radiant heat.
Answer C is wrong. The thermometer should be placed four inches from the surface of the container.
**Answer D is correct.** When checking a refrigerant container for noncondensable gases, the container should be stored out of the presence of sunlight at 65° F (18° C) for 12 hours, and a thermometer should be placed 4 inches (102 mm) from the container surface. If the container pressure is less than specified, the refrigerant is ready for use.

Question #78

**Answer A is correct.** A refrigerant identifier analyzes the refrigerant to determine what is mixed with it. Pure R-12 and R-134a can be identified by their chemical content. Other chemicals mixed with the refrigerant or mixed R-12 and R134a can be identified. These other chemicals include air, flammables such as propane, and blended refrigerants.

Answer B is wrong. The identifier cannot detect an excessive charge. The identifier is able to detect the percentages of R12, R134a, air, flammable refrigerants and blended refrigerants.

Answer C is wrong. The identifier cannot detect excessive oil in the system. The identifier is able to detect the percentages of R12, R134a, air, flammable refrigerants and blended refrigerants.

Answer D is wrong. The identifier cannot detect a low refrigerant charge. The identifier is able to detect the percentages of R12, R134a, air, flammable refrigerants and blended refrigerants.

# Answers to the Test Questions for the Sample Test Section 6

| | | | | | | | |
|---|---|---|---|---|---|---|---|
| 1. | C | 19. | C | 38. | A | 56. | B |
| 2. | D | 20. | B | 39. | C | 57. | D |
| 3. | D | 21. | B | 40. | C | 58. | C |
| 4. | C | 22. | C | 41. | B | 59. | A |
| 5. | C | 23. | B | 42. | C | 60. | C |
| 6. | D | 24. | D | 43. | C | 61. | B |
| 7. | C | 25. | A | 44. | A | 62. | B |
| 8. | C | 26. | C | 45. | A | 63. | D |
| 9. | C | 27. | C | 46. | C | 64. | C |
| 10. | C | 28. | A | 47. | C | 65. | C |
| 11. | C | 29. | A | 48. | B | 66. | C |
| 12. | C | 30. | A | 49. | A | 67. | C |
| 13. | A | 31. | A | 50. | C | 68. | C |
| 14. | C | 32. | C | 51. | C | 69. | C |
| 15. | C | 33. | C | 52. | C | 70. | A |
| 16. | C | 34. | B | 53. | B | 71. | B |
| 17. | A | 35. | A | 54. | D | 72. | A |
| 18. | C | 36. | B | 55. | B | 73. | D |
| | | 37. | A | | | | |

# Explanations to the Answers for the Sample Test Section 6

Question #1
Answer A is wrong. Technician B is also correct.
Answer B is wrong. Technician A is also correct.
**Answer C is correct.** Both technicians are correct. The blower motor voltage drop in the high-speed position should be near system voltage to provide high speed. The cause of the problem could be as a result of low voltage caused by high resistance or low source voltage supply. A high resistance fault at the ground connection could cause some of the voltage to be dropped at the ground circuit. Voltage drop testing is a common way to isolate resistance problems.
Answer D is wrong. Both technicians are correct.

Question #2
Answer A is wrong. The high pressure hose should be connected to a point in the high side of the system. The high side of the system runs from the compressor discharge hose to the metering device.
Answer B is wrong. The low pressure hose should be connected to a point in the low side of the system. The low side of the system runs from the metering device to the suction line at the compressor.
Answer C is wrong. Neither technician is correct.
**Answer D is correct.** Neither technician is correct. The high-pressure hose should be connected at any point between the outlet of the compressor and the expansion device separating the high and low side. The most common point is at the compressor outlet or near it. The low-pressure hose should be connected to the low-pressure hose running from the evaporator to compressor. The most common point on an orifice tube system would be on the accumulator.

Question #3
Answer A is wrong. Charging the A/C system into the high side must never be done with the system turned on. Doing so could cause high pressure to be pumped from the vehicle into the charging equipment.
Answer B is wrong. Having the system in a vacuum is the recommended practice when manually charging the A/C system. This allows for better initial transfer of refrigerant from the tank to the vehicle.
Answer C is wrong. Inverting a tank of refrigerant causes liquid to be extracted during the charge process.
**Answer D is correct.** The A/C system should only be charged through the high side with the system off. When the system is in a vacuum the pressure of the refrigerant will push the refrigerant into the system. Inverting the refrigerant container disperses liquid refrigerant. By charging with liquid, the pressure of the refrigerant container doesn't drop as rapidly. As the system is charged, the system pressures should equalize.

Question #4
Answer A is wrong. Technician B is also correct.
Answer B is wrong. Technician A is also correct.
**Answer C is correct.** Both technicians are correct. When the screen of an orifice tube is restricted, the technician needs to determine the cause. Metallic particles will usually come from the compressor and indicate the potential of a failing compressor. When the orifice tube is restricted, it causes a reduced amount of refrigerant volume to the evaporator and a resulting pressure drop. This may result in frost forming at the point of restriction.
Answer D is wrong. Both technicians are correct.

Question #5
Answer A is wrong. Technician B is also correct. a good choice but not the correct answer because both technicians are correct.
Answer B is wrong. Technician A is also correct. a good choice but not the correct answer because both technicians are correct.
**Answer C is correct.** Both technicians are correct. Converting a R-12 system to use R-134A can be done if all federal and local regulations are met. Typically the R-12 system must be thoroughly purged and flushed to remove all traces of R-12. Different systems require that different components be changed as well. It is recommended that as much R-12 system oil as possible be removed from the system. An R-134a-compatible oil should be used in the proper amount. After the conversion has been completed, the system must be labeled as an R-134A system. All standard or required service procedures (such as reclaiming) must be followed during conversion.
Answer D is wrong. Both technicians are correct.

Question #6
Answer A is wrong. A slipping fan clutch would not affect ram airflow. This problem would reduce airflow at low speeds causing reduced air conditioning performance at low speeds.
Answer B is wrong. Reduced airflow across the condenser would elevate both the high and low side pressures.
Answer C is wrong. Neither technician A nor B is correct.
**Answer D is correct.** Neither technician is correct. Airflow across the condenser is created by vehicle movement. A fan clutch allows the fan to freewheel when ram airflow is greater than the fan induced airflow. A slipping clutch would not change ram airflow. It is also common for a fan clutch to lock-up which will cause problems for the water pump, the belt drive, and the engine's horsepower output. A locked up fan clutch will not reduce A/C cooling efficiency. Reduced airflow across the condenser generally affects the ability of the condenser to cool the refrigerant, resulting in higher temperatures and pressures in both sides of the A/C system.

Question #7
Answer A is wrong. Technician B is also correct.
Answer B is wrong. Technician A is also correct.
**Answer C is correct.** Both technicians are correct. When retrofitting an A/C system, it is necessary to replace the system O-rings on most systems while retrofitting because the seals used in R-134A systems are silicon based. Even if there are not leaks in the system before retrofitting there may be shortly after. To avoid time consuming comebacks, it is the best practice to replace all seals during the original retrofit. Leak testing the system is one of the first steps that are taken while converting a R-12 system to R-134A. The accumulation of dirt around an A/C line connection is an indication of a leak in the system. When there is a leak in the system, dirt collects in the area of the leak due to some of the oil leaking out.
Answer D is wrong. Both technicians are correct.

Question #8
Answer A is wrong. Technician B is also correct.
Answer B is wrong. Technician A is also correct.
**Answer C is correct.** Both technicians are correct. If the condenser air passages are severely restricted, this may result in high refrigerant system pressures and refrigerant discharge from the high-pressure relief valve. These restrictions will not allow enough air to cool down the condenser, causing higher temperatures and result in higher pressures.
Answer D is wrong. Both technicians are correct.

Question #9
Answer A is wrong. Technician B is also correct.
Answer B is wrong. Technician A is also correct.
**Answer C is correct.** Both technicians are correct. When the engine is running with the thermostat closed, the coolant is directed through a bypass system to be circulated again through the engine and heater core. The heater core is part of the bypass system on some vehicles.
Answer D is wrong. Both technicians are correct.

Question #10
Answer A is wrong. Technician B is also correct.
Answer B is wrong. Technician A is also correct.
**Answer C is correct.** Both technicians are correct. All of the refrigerant in the system must be recovered prior to parts replacement or evacuation. Evacuation is the name given to the process that pulls all traces of air, moisture, and refrigerant from the system. This is done by creating a vacuum in the system. A vacuum pump is connected to the system and should remain on and connected to the system for at least 30 minutes after 20 to 29 inches of mercury is reached. Any air or moisture that is left inside an air conditioning system reduces the system's efficiency and eventually leads to major problems, such as compressor failure. Air causes excessive pressure within the system, restricting the refrigerant's ability to change its state from gas to liquid within the refrigeration cycle, which drastically reduces its heat absorbing and transferring ability. Moisture, on the other hand, can cause freeze-up at the cap tube or expansion valve, which restricts refrigerant flow or blocks it completely. Both of these problems result in intermittent cooling or no cooling at all. Moisture also forms hydrochloric acid when mixed with refrigerant, causing internal corrosion, which is especially dangerous to the compressor.
Answer D is wrong. Both technicians are correct.

Question #11
Answer A is wrong. Technician B is also correct.
Answer B is wrong. Technician A is also correct.
**Answer C is correct.** Both technicians are correct. Some high-pressure relief devices are spring-loaded and self-resetting. Some high-pressure relief devices are made of fusible metal and are not self-resetting. Self-resetting high-pressure relief valves are usually found on the rear head of the compressor while fusible metal high-pressure relief devices are often found on the receiver/drier, or some other high-side component.
Answer D is wrong. Both technicians are correct.

Question #12
Answer A is wrong. Technician B is also correct.
Answer B is wrong. Technician A is also correct.
**Answer C is correct.** Both technicians are correct. A pressure cycling switch is electrically connected in series with the compressor electromagnetic clutch. Like the thermostatic switch, the turning on and off of the pressure cycling switch controls the operation of the compressor. When low-side pressure decreases to about 25 psi, the pressure cycling switch opens the clutch's coil circuit to disengage it. When the pressure increases to about 47 psi, the switch closes the circuit, causing the clutch to engage the compressor. Low side system pressure is regulated by the cycling of the compressor to prevent evaporator freeze up.
Answer D is wrong. Both technicians are correct.

Question #13
**Answer A is correct.** Only Technician A is correct. A compressor noise problem is often difficult to locate. When the compressor is operating, a loose compressor mount or bracket may vibrate, giving the impression that the compressor itself is somehow defective. This type of noise usually stops when the compressor is disengaged.
Answer B is wrong. If the system did not have any refrigerant in it, then the low pressure devices would prevent the A/C compressor from running.
Answer C is wrong. Only Technician A is correct.
Answer D is wrong Only Technician A is correct.

Question #14
Answer A is wrong. Technician B is also correct.
Answer B is wrong. Technician A is also correct.
**Answer C is correct.** Both technicians are correct. When converting to R-134A, the rules really depend on the vehicle. There are very few absolutes except that the oil must be changed. Some vehicles will rapidly fail if their original compressor is utilized. Others are fine to convert. Most systems require changes to nitril seals and special coated gaskets. There are a few applications that used nitril seals in the OE R-12 application. The key here is to be sure that you check about compatibility of components. There are some systems that cannot be successfully converted.
Answer D is wrong. Both technicians are correct.

Question #15
Answer A is wrong. Technician B is also correct.
Answer B is wrong. Technician A is also correct.
**Answer C is correct.** Both technicians are correct. A leaking heater core is often the cause of a windshield fogging problem. Windshield fogging is the result of hot and humid air that condenses on the cooler glass. Under this condition, the coolant level in the radiator should be checked. The heater core should be repaired or replaced. Technician B is also right. Cleaning a clogged evaporator/heater case drain tube often eliminates a windshield fogging problem. If the drain is clogged, a leaking heater core will allow engine coolant to accumulate in the housing. A clogged drain also allows condensed water vapor to accumulate. Both of these problems add moisture to the air, which tends to fog the windshield.
Answer D is wrong. Both technicians are correct.

Question #16
Answer A is wrong. Technician B is also correct.
Answer B is wrong. Technician A is also correct.
**Answer C is correct.** Both technicians are correct. This is a legal question and the answer depends on your knowledge of federal, state, and local laws. Technician A is right. Coolant is considered a hazardous waste and must be recovered. Technician B is right. Coolant can be recycled according to local regulations. In no instance can it be disposed of by pouring on the ground or in a landfill. Any coolant removed from an automotive cooling system may contain harmful engine residues, such as benzene and lead. If used coolant is to be disposed of, it must be done according to local regulations. Ideally, however, the coolant may be recycled by filtering out most of the contaminants. Some recycle machines use vacuum distillation or ion exchange technology for glycol-based coolants. After recycling, the coolant must be supplemented with an additive before being reused.
Answer D is wrong. Both technicians are correct.

Question #17
**Answer A is correct.** Only Technician A is correct. The high-side pressure cutoff switch is wired with the compressor clutch (in series). Designed to open (cut out) and disengage the clutch at 350 to 375 psi, it again closes and normally reengages the clutch when pressure returns to 250 psi (higher if the system uses R-134A). This prevents the venting of refrigerant into the atmosphere by the pressure relief valve. The high-pressure relief valve is a high-pressure protection device that opens (normally at 440 psi) to bleed off excessive pressure that might occur in the system. The relief valve is a simple poppet-type valve in the compressor and is not part of the electrical circuit.
Answer B is wrong. The pressure relief valve mechanically vents refrigerant if pressures rise above 440 psi to prevent component damage from occurring.
Answer C is wrong. Only Technician A is correct.
Answer D is wrong. Only Technician A is correct.

Question #18

Answer A is wrong. A leaking desiccant bag would cause dark pieces of desiccant to travel through the system.

Answer B is wrong. A plugged condenser would not cause aluminum particles to end up at the orifice tube.

**Answer C is correct.** The only possible source for aluminum is the compressor. If the filter of an orifice tube is covered with aluminum particles, the aluminum must be from the compressor. The compressor must be damaged. The other answer choices would not release aluminum particles into the refrigerant.

Answer D is wrong. The evaporator is after the orifice tube in refrigeration circuit and would not cause particles to end up at the orifice tube.

Question #19

Answer A is wrong. An over-charge of refrigerant will cause excessive system pressures.

Answer B is wrong. Restricted airflow at the condenser will cause excessive system pressures because the hot vapor will not be able to give up its heat.

**Answer C is correct.** Interpreting manifold gauge readings is a crucial part of evaluating and diagnosing an A/C system. What follows is a quick summary of what to suspect when the readings are abnormal. If the high-side pressure is too high, suspect air in the system, too much refrigerant in the system, a restriction in the high side of the system, and poor airflow across the condenser (in this question caused by the slipping fan clutch). If the high-side pressure is too low, suspect a low refrigerant level or defective compressor. If the low-side pressure is higher than normal, suspect refrigerant overcharge, a defective compressor, or a faulty metering device. If the low-side pressure is lower than normal, suspect a faulty metering device, poor airflow across the evaporator, a restriction in the low side of the system, or the system is undercharged with refrigerant. Answer C is the EXCEPT answer because poor airflow across the evaporator would lower the high-side pressure because little heat is added to the refrigerant.

Answer D is wrong. A slipping fan clutch will cause excessive system pressures because airflow through the condenser would be affected.

Question #20

Answer A is wrong. A broken capillary tube will cause poor cooling performance.

**Answer B is correct.** The only probable result of a broken expansion valve capillary tube would be inadequate cooling. A defective or broken capillary tube would Most-Likely cause the thermal expansion valve (TXV) to remain closed, causing poor cooling at all times. Therefore the other choices are problems, but these problems would not be caused by a broken or defective capillary tube.

Answer C is wrong. Charging the A/C system would not be affected by a broken capillary tube on the TXV.

Answer D is wrong. The air conditioner performance would surely be negatively affected with a broken capillary tube on the TXV.

Question #21

Answer A is wrong. A loose belt will usually only make a squealing noise when the clutch is engaged.

**Answer B is correct.** Only Technician B is correct. A serpentine belt is kept at the proper tension by an automatic tensioner. When the tensioner can't maintain the correct pressure the belt may slip and possibly squeal when the compressor clutch engages. Most automatic tensioners can indicate when the belt is stretched enough to require replacement. A loose A/C compressor belt will likely produce a squealing noise only when the compressor clutch is engaged. A bad compressor pulley bearing could cause the belt to squeal all of the time.

Answer C is wrong. Only Technician B is correct.

Answer D is wrong. Only Technician B is correct.

Question #22
Answer A is wrong. An undercharged system would produce lower system pressures than the ones described in this question.
Answer B is wrong. An overcharged system would produce higher system pressures than the ones described in this question.
**Answer C is correct.** The gauge readings stated in this question are normal for most systems, especially for a system with a cycling clutch and a pressure switch to cycle the compressor. Low refrigerant level or an undercharge condition would cause the readings to be low. An overcharge condition would result in higher-than-normal pressures. And if the evaporator pressure regulator were bad, the high-side pressure would be low.
Answer D is wrong. A bad evaporator pressure regulator would cause lower than normal high-side pressures.

Question #23
Answer A is wrong. Answer A is a true statement but is not the correct answer because this is an EXCEPT question.
**Answer B is correct.** The components that are numbered in the drawing for this question are: 1, the compressor, 2, the condenser, 3, the evaporator, 4, the accumulator, and 5, some type of switch located on the accumulator. Any type of low-pressure switch could be located there. The high-pressure cutoff switch is a high-pressure switch that keeps system pressures from reaching a point that may cause compressor lockup or other component damage because of excessive high pressures. Because it is a high-pressure switch it is the EXCEPT answer and B is the proper choice. All other switches are low-pressure devices.
Answer C is wrong. Answer C is a true statement but is not the correct answer because this is an EXCEPT question.
Answer D is wrong. Answer D is a true statement but is not the correct answer because this is an EXCEPT question.

Question #24
Answer A is wrong. The cause of this problem could be an open compressor clutch coil.
Answer B is wrong. The cause of this problem could be a shorted compressor clutch coil.
Answer C is wrong. The cause of this problem could be an open compressor clutch circuit ground.
**Answer D is correct.** The control circuit for the compressor clutch is being bypassed and it still does not work. The question asks for the LEAST-Likely cause. An open wire in the compressor clutch control circuit would not cause this problem. By using the jumper wire, the control circuit is being bypassed.

Question #25
**Answer A is correct.** Only Technician A is correct. The power steering pressure switch signals the computer to increase the engine idle speed during parking-type maneuvers. If the vehicle has a bad power steering pressure switch, the engine could stall at idle when the steering is moved to the full lock position. This is due to the increased load on the engine from the power steering pump.
Answer B is wrong. A loose belt would not cause the engine to die during this situation.
Answer C is wrong. Only Technician A is correct.
Answer D is wrong. Only Technician A is correct.

Question #26

Answer A is wrong. Technician B is also correct.

Answer B is wrong. Technician A is also correct.

**Answer C is correct.** Both technicians are correct. Testing the refrigerant system for leaks is one of the most important phases of troubleshooting. Over a period of time, all air conditioners lose or leak some refrigerant. Leaks are most often found at the compressor hose connections and at the various fittings and joints in the system. Refrigerant can be lost through hose permeation. Leaks can also be traced to pinholes in the evaporator caused by acid, which forms when water and refrigerant mix. Since oil and refrigerant leak out together, oily spots on hoses, fittings, and components means there is a leak. Using an electronic leak detector is the preferred method of leak detection; it is safe, effective, and can be used with all types of refrigerants. The hand-held battery-operated electronic leak detector contains a test probe that is moved about one inch per second in areas of suspected leaks. (Remember that refrigerant gas is heavier than air, thus the probe should be positioned below the test point.) An alarm or a buzzer on the detector indicates the presence of a leak. On some models, a light flashes to establish the leak. The electronic leak detector is the more sensitive of the two types. Other methods of leak detection include fluorescent leak tracers and fluid-type leak detectors. All leaks should be repaired before refrigerant is added to the system.

Answer D is wrong. Both technicians are correct.

Question #27

Answer A is wrong. Technician B is also correct.

Answer B is wrong. Technician A is also correct.

**Answer C is correct.** Both technicians are correct. Restricted radiator tubes may cause engine overheating. The engine overheats and causes excessive cooling system temperatures and pressures, forcing excessive coolant flow into the recovery container. As the system cools, it creates a vacuum and draws the coolant from the recovery container back into the radiator (cooling system). A blown head gasket could also cause the engine to overheat and push coolant into the recovery tank. If the coolant wasn't allowed to leak out onto the ground or if it was not leaking into the combustion chamber, then the level could return back to the normal level.

Answer D is wrong. Both technicians are correct.

Question #28

**Answer A is correct.** Only Technician A is correct. An improperly positioned or missing radiator fan shroud could cause overheating. The radiator shroud needs to be properly positioned to aid the fan in directing airflow through the radiator. Technician B is wrong. The fan clutch being locked would not cause overheating. A locked fan clutch would cause the fan to run at all times. This would cause an overcooling, if anything. The biggest problem with a locked fan clutch is the increased drag on the engine costing fuel and horsepower.

Answer B is wrong. A seized thermostatic fan clutch would not cause the engine to overheat when parked. The clutch being seized means that it is running at full speed all of the time; which would not cause an overheating condition.

Answer C is wrong. Only Technician A is correct.

Answer D is wrong. Only Technician A is correct.

Question #29

**Answer A is correct.** Only Technician A is correct. Whenever a fuse blows, it is caused by excessive current flow. In this case, the most probable cause of a blown fuse is a short in the circuit or a binding motor. An open field winding in the motor would not cause an increase in current flow and therefore would not blow the fuse. If a short is suspected, the motor should be checked first, then the circuit. If the motor is defective, replace it. If the circuit is shorted, repair it and then replace the fuse.

Answer B is wrong. An open field winding would likely cause the blower to be inoperative but it would not cause high current to flow and open up the circuit protection device.

Answer C is wrong Only Technician A is correct.

Answer D is wrong Only Technician A is correct.

Question #30
**Answer A is correct.** Only Technician A is correct. When the fins and air passages of an evaporator are clogged with dirt, air will have a difficult time passing through the evaporator. Since the principles of air conditioning require the heated air around the evaporator to lose heat, a vehicle with a clogged evaporator will have poor A/C performance. Technician A is right in saying an evaporator core should be removed from the case to properly clean it. Technician B is not right. Airborne debris, such as leaves, lint, hair, and other contaminants, enters the evaporator case and clings to the wet evaporator core. The drain is for the purpose of removing the condensation that forms as a result of humidity being removed from the intake air. It is nearly impossible to adequately clean an evaporator core without first removing it from the case.
Answer B is wrong. Airborne debris, such as leaves, lint, hair, and other contaminants, enters the evaporator case and clings to the wet evaporator core. The drain is for the purpose of removing the condensation that forms as a result of humidity being removed from the intake air.
Answer C is wrong. Only Technician A is correct.
Answer D is wrong. Only Technician A is correct.

Question #31
**Answer A is correct.** Typically the causes for excessive high-side pressures are air in the system, too much refrigerant in the system, a restriction in the high side of the system, and poor airflow across the condenser.
Answer B is wrong. A leaking thermostatic expansion valve would cause the system to have low operating pressures.
Answer C is wrong. A bypass valve is not part of a typical A/C system.
Answer D is wrong. A vapor control switch is not part of a typical A/C system.

Question #32
Answer A is wrong. Technician B is also correct.
Answer B is wrong. Technician A is also correct.
**Answer C is correct.** Both technicians are correct. Performance testing provides a measure of air conditioning system operating efficiency. A manifold pressure gauge set is used to determine both high and low pressures in the refrigeration system. The desired pressure readings will vary according to temperature. Use temperature/ pressure charts as a guide to determine the proper pressures. At the same time, a thermometer is used to determine air discharge temperature into the passenger compartment. The typical procedure for a performance test follows: Carefully look over all of the components and lines of the system. Connect the manifold gauge set to the respective high- and low-pressure fittings. These fittings are found in various locations within the high- and low-pressure sides of the system. Close the hood and all of the door and windows of the vehicle. Adjust the air conditioning controls to maximum cooling and high blower position. Idle the engine in neutral or park with the brake on. For the best results, place a high volume fan in front of the radiator grille to insure an adequate supply of airflow across the condenser. Increase engine speed to 1500 to 2000 rpm. Measure the temperature at the evaporator air outlet grille or air duct nozzle (35 to 40 ° F). Read the high and low pressures and compare them to the normal range of the operating pressure given in the service manual. Both technicians identified steps for a valid performance test..
Answer D is wrong. Both technicians are correct.

Question #33
Answer A is wrong. Technician B is also correct.
Answer B is wrong. Technician A is also correct.
**Answer C is correct.** Both technicians are correct. The control panel assembly controls the compressor, heater valve, and plenum door. The control panel assembly has no control over the coolant temperature. The purpose of the control panel is to provide operator input for the air conditioning and heating system. Some control panels have provisions for displaying in-car and outside air temperatures. A microprocessor may be located in the control panel to provide input data to the programmer, based on operator-selected conditions.
Answer D is wrong. Both technicians are correct.

Question #34

Answer A is wrong. A technician should never charge into the high-side with the air conditioning turned on.
**Answer B is correct.** Only Technician B is correct. If liquid refrigerant enters the compressor, the compressor can be damaged. Technician A is supporting a very dangerous practice. If you connect a refrigerant can to the high side with the engine running and the system in operation, the high pressure from the system can flow back into the can and cause it to burst. The high-side charging procedure must only be completed with the engine off. The vehicle should always be charged from the low-pressure side if the system is in operation. High-side charging with the system off is accomplished by putting the system into a vacuum and using the refrigerant pressure to push the charge into the system. Heat can also be applied to the charge cylinder to cause its pressure to be above that of the system.
Answer C is wrong. Only Technician B is correct.
Answer D is wrong. Only Technician B is correct.

Question #35

**Answer A is correct.** In cycling clutch systems with a fixed orifice tube, a low-pressure or cycling clutch switch is placed in series with the compressor clutch circuit so it can turn the clutch on or off. It de-energizes the clutch and stops the compressor if the pressure gets below a certain level, (usually about 20-25 psi). It closes and turns on the compressor at about 40 psi.
Answer B is wrong. The condenser is in the high-side of the system and would not need any type of de-icing device.
Answer C is wrong. Only Technician A is correct.
Answer D is wrong. Only Technician A is correct.

Question #36

Answer A is wrong. The question describes testing for source vacuum. It does not describe any results from tests performed on the mode door.
**Answer B is correct.** The vacuum source should be equal to engine vacuum and a blocked, disconnected, or kinked line could cause the zero vacuum reading.
Answer C is wrong. The question describes testing for source vacuum. It does not describe any results from tests performed on the vacuum motor.
Answer D is wrong. The question describes testing for source vacuum. If no vacuum was coming from the engine, then the motor is not going to be operational.

Question #37

**Answer A is correct.** Evacuation is the name for the process of connecting a vacuum pump and allowing it to operate for an amount of time dictated by system conditions. Proper evacuation rids the system of unwanted air and moisture that recovery did not remove. Air is pumped out by the vacuum pump as soon as it reaches a vacuum. The moisture is removed by creating a vacuum that will cause any moisture to be boiled into a vapor that can be pumped out by the vacuum. This process can take up to 4 hours depending on the level of contamination, system temperature, and ability of the vacuum pump. Rust, dirt, desiccant, and anything else that was not a vapor would probably not be pulled through the opening of the service valves. If these materials were present in the refrigerant, the system would have problems and parts would need to be replaced.
Answer B is wrong. Rust could only be removed by flushing or replacing system components.
Answer C is wrong. Dirt could only be removed by flushing or replacing system components.
Answer D is wrong. Desiccant particles could only be removed by flushing or replacing system components.

Question #38

**Answer A is correct.** Only Technician A is correct. Typically, problems with the heating system are problems with the engine's cooling system. Problems that pertain specifically to the heater are few: the heater control valve and the heater. When there is a problem of insufficient heat, begin your diagnosis with a visual inspection and a check of the coolant level. If the level is correct, turn the heater controls on and run the engine until it reaches normal operating temperature. Then measure the temperature of the upper radiator hose. The temperature can be measured with a pyrometer. If one is not available, gently touch the hose. You should not be able to hold the hose long because of the heat. While doing this, make sure you stay clear of the area around the cooling fan. A spinning fan can chop off your hand. If the temperature of the hose is not within specifications, suspect a faulty thermostat. If the hose is the correct temperature, check the temperature of the two heater hoses. They should both be hot. If only one of the hoses is hot, suspect the heater control valve or a plugged heater core.
Answer B is wrong. If the cooling fans do not work, then the vehicle will overheat when driven at low speeds or when it is allowed to idle for long periods of time.
Answer C is wrong. Only Technician A is correct.
Answer D is wrong. Only Technician A is correct.

Question #39

Answer A is wrong. Technician B is also correct.
Answer B is wrong. Technician A is also correct.
**Answer C is correct.** Both technicians are correct. If the heater core is restricted, coolant flow may be prevented. Any condition that will affect the proper flow of engine coolant through the heater core may affect proper heater performance. This includes a defective control valve or a blockage in the heater hoses or core. The thermostat's primary purpose is to prevent engine overcooling. A stuck-open thermostat will delay engine warm-up and may reduce available heat to the heater core.
Answer D is wrong. Both technicians are correct.

Question #40

Answer A is wrong. Technician B is also correct.
Answer B is wrong. Technician A is also correct.
**Answer C is correct.** Both technicians are correct. Some compressor clutch circuits contain a thermal limiter switch that senses compressor surface temperature. Some A/C compressor clutch circuits also contain a low-pressure cutoff switch. A/C systems have load-sensitive electrical Switches; such as the low-pressure switch, high-pressure switch, pressure cycling, and power steering. Not all of these switches, however, are used on all vehicles. The pressure switches are used to prevent compressor or component damage in the event of extremely high- or low-system pressures.
Answer D is wrong. Both technicians are correct.

Question #41

Answer A is wrong. A refrigerant leak doesn't typically make a noise, except when the refrigerant blows out of the leak. This noise would happen suddenly and would not have a hissing sound.
**Answer B is correct.** Only Technician B is correct. On most systems, when the engine is turned off after the A/C system has been on, a hissing sound will be heard as the pressure within the system equalizes. This is a normal noise. A refrigerant leak doesn't typically make a noise, except when the refrigerant blows out of the leak. This noise would happen suddenly and would not have a hissing sound.
Answer C is wrong. Only Technician B is correct.
Answer D is wrong. Only Technician B is correct.

Question #42

Answer A is wrong Technician B is also correct.

Answer B is wrong. Technician A is also correct.

**Answer C is correct.** Both technicians are correct. Refrigerant oil is as moisture-free as it is possible to make it. It is called a hygroscopic material because it quickly absorbs water or moisture. For this reason, the oil container should not be opened until ready for use and then it should be capped immediately after use. All air contains moisture. Air that enters any part of the system carries moisture with it, and the exposed surface collects the moisture quickly. To remove the moisture and air, an A/C system must be evacuated (vacuumed) any time it has a leak or the system is exposed to air during service.

Answer D is wrong. Both technicians are correct.

Question #43

Answer A is wrong. Technician B is also correct.

Answer B is wrong. Technician A is also correct.

**Answer C is correct.** Both technicians are correct. If the high-side pressure is too high, suspect air in the system, too much refrigerant in the system, a restriction in the high side of the system, and poor airflow across the condenser. If the low-side pressure is higher than normal, suspect refrigerant overcharge, a defective compressor, or a faulty metering device. When both pressures are higher than normal, suspect poor airflow across the condenser, an overheating engine, or an overcharge condition. Good airflow through the condenser is necessary to remove heat from the refrigerant. Often this airflow is restricted by a buildup of dirt or bugs on the outside of the condenser. Not only will a dirty condenser cause insufficient cooling, it can also cause the engine to overheat, as the airflow across the radiator will also be reduced.

Answer D is wrong. Both technicians are correct.

Question #44

**Answer A is correct.** Only Technician A is correct. Air conditioning systems are extremely sensitive to moisture and dirt. Therefore, clean working conditions are extremely important. The smallest particle of foreign matter in an air conditioning system contaminates the refrigerant, causing rust, ice, or damage to the compressor. For this reason, all replacement parts are sold in vacuum-sealed containers and should not be opened until they are to be installed in the system. If, for any reason, a part has been removed from its container for any length of time, the part must be completely flushed using only recommended flush solvent to remove any dust or moisture that might have accumulated during storage. When the system has been open for any length of time, the entire system must be purged completely, and a new receiver/drier must be installed because the element of the existing unit will have become saturated and unable to remove any moisture from the system once the system is recharged. Technician A is the correct choice; just one drop of water added to the refrigerant will start chemical changes that can result in corrosion and eventual breakdown of the chemicals in the system. Hydrochloric acid is the result of an R-12 mixture with water. The smallest amount of moist air in the refrigerant system might start reactions that can cause malfunctions.

Answer B is wrong. It is normal for refrigerant oil to travel through the A/C system as the refrigerant is circulated through the system.

Answer C is wrong. Only Technician A is correct.

Answer D is wrong. Only Technician A is correct.

Question #45

**Answer A is correct.** Only Technician A is correct. The receiver/drier is a storage tank for the liquid refrigerant from the condenser. It also contains a bag of desiccant. The purpose of the desiccant is to absorb any moisture that might enter the system during assembly or service. The receiver/drier is often neglected when the air conditioning system is serviced or repaired. Failure to replace it can lead to poor system performance or replacement part failure. It is recommended that the receiver/drier and/or its desiccant be changed whenever a component is replaced, the system has lost the refrigerant charge, or the system has been open to the atmosphere for any length of time. If the receiver/drier outlet is colder than the inlet, this unit is restricted and should be changed to restore efficient and proper system operation.

Answer B is wrong. If the refrigerant in the sight glass appears red, leak-detecting dye has been added to the refrigerant. This condition does not require any corrective action.

Answer C is wrong. Only Technician A is correct.

Answer D is wrong. Only Technician A is correct.

Question #46
Answer A is wrong. Technician B is also correct.
Answer B is wrong. Technician A is also correct.
**Answer C is correct.** Both technicians are correct. The purpose of a typical automotive heater/air conditioner/case and duct system is twofold. It is used to house the heater core and the air conditioner evaporator and to direct the selected supply air through these components into the passenger compartment of the vehicle. The supply air selected can be either outside or re-circulated air, depending on the system mode. After the air is heated or cooled, it is delivered to the floor outlet, dash panel outlets, or the defrost outlets. All incoming air is directed through the evaporator core in all modes of the climate control system using the evaporator case, evaporator case doors, and the blower motor. The purpose of having all air routed through the evaporator is to remove humidity from the air during the defroster operation.
Answer D is wrong. Both technicians are correct.

Question #47
Answer A is wrong. The suction line connects the evaporator to the inlet of the compressor.
Answer B is wrong. Suction lines carry low temp/low pressure vapor from the evaporator to the compressor.
**Answer C is correct.** Suction lines can be distinguished from the discharge lines by touch and size. They are *cold* to the touch, not hot as indicated by statement C. Discharge (high-pressure) and liquid lines are very warm to the touch and easily distinguishable from the suction lines.
Answer D is wrong. Vapor takes up more space that liquid refrigerant.

Question #48
Answer A is wrong. Air conditioning equipment service hoses must have shutoff valves within 12 inches of the service ends.
**Answer B is correct.** To minimize the amount of refrigerant released to the atmosphere when A/C systems are serviced, always follow these steps. The recycling equipment must have shutoff valves within 12 inches of the hoses' service ends. With the valves closed, connect the hoses to the vehicle's air conditioning service fittings. Recover the refrigerant from the vehicle and continue the process until the vehicle's system shows vacuum instead of pressure. Turn off the recovery/recycling unit for at least five minutes. If the system still has pressure, repeat the recovery process to remove any remaining refrigerant. Continue until the A/C system holds a stable vacuum for two minutes. Close the valves in the recovery/recycling unit's service lines and disconnect them from the system's service fittings.
Answer C is wrong. Air conditioning systems should be recovered until all positive pressure is removed the system.
Answer D is wrong. A wise technician waits five minutes after the recovery is complete to check for positive pressure in the air conditioning system.

Question #49
**Answer A is correct.** The oil required with R-134A refrigerant is a polyalkylene glycol (PAG) oil. This synthetic oil is very hygroscopic and will quickly absorb any moisture that it comes in contact with.
Answer B is wrong. Engine oil should never be used in an AC system.
Answer C is wrong. Mineral oil is petroleum based and synthetic oil is made by man.
Answer D is wrong. Polyalkylene glycol (PAG) is a synthetic oil which means it is made by man.

Question #50
Answer A is wrong. Technician B is also correct.
Answer B is wrong. Technician A is also correct.
**Answer C is correct.** Both technicians are correct. Vacuum motors move the various doors to direct the airflow to the proper vents. Vacuum air doors are controlled by a selector switch in the control head (panel). This unit directs the vacuum signal to the proper motor. If the selector switch is defective, vacuum will not be supplied to the proper door. If there is a leak in the vacuum system, the door will not move properly due to no or a low vacuum signal. A vacuum leak anywhere in the control circuit will affect the operation of the entire system. If the system is sound, the vacuum motor should be checked. When testing a vacuum actuated door, the vacuum gauge reading should remain steady for at least one minute. If the gauge reading drops slowly, the actuator is leaking and should be replaced.
Answer D is wrong. Both technicians are correct.

Question #51
Answer A is wrong. Technician B is also correct.
Answer B is wrong. Technician A is also correct.
**Answer C is correct** Both technicians are correct. The capillary tube controls the amount the expansion valve is open. As the temperature of the outlet of the evaporator increases, the expansion valve opens to allow for more refrigerant flow. The temperature is sensed by a remote thermal bulb attached to the capillary tube of the valve. A decrease in outlet temperature will cause the valve to close, thereby restricting the flow of refrigerant. A quick test of the expansion valve involves warming and cooling the thermal bulb and watching the change in the pressures on the manifold gauge set.
Answer D is wrong. Both technicians are correct.

Question #52
Answer A is wrong. Technician B is also correct.
Answer B is wrong. Technician A is also correct.
**Answer C is correct.** Both technicians are correct. If there are 12 volts at the compressor clutch, there is normal battery voltage supplied to the compressor clutch and connecting wire. If the clutch is inoperative, the compressor clutch coil or the connection from the coil to ground is open.
Answer D is wrong. Both technicians are correct.

Question #53
Answer A is wrong. The water control valve should be completely open with the temperature control on hot.
**Answer B is correct.** A sticking or disconnected blend door will affect temperature of the air but will not affect the point of air discharge.
Answer C is wrong. Vacuum should be present at N.C. heater control valve when the control head is set to heat.
Answer D is wrong. Air pockets in the heater core negatively affect heater output.

Question #54
Answer A is wrong. It is possible that a vehicle could have had refrigerants other than R12 installed into its A/C system. A wise technician always uses an identifier before hooking shop equipment up to a vehicle.
Answer B is wrong. It is possible that a vehicle could have had refrigerants other than R12 installed into its A/C system. A wise technician always uses an identifier before hooking shop equipment up to a vehicle.
Answer C is wrong. Neither technician is correct.
**Answer D is correct.** Neither technician is correct. It must never be assumed that a system is charged with R-12 simply because it has 1/4-inch service valves. It should never be assumed that the lack of a label or labels is an indication that the system is charged with R-12 refrigerant. Other refrigerants have been used in automotive air conditioning systems without regard to product identification. This may even include some refrigerants that are considered flammable or otherwise hazardous and not approved for automotive service. It, therefore, should never be assumed that a system is charged with any particular type of refrigerant simply because of its service valves or the lack of labels. A refrigerant identifier has become a necessary service tool.

Question #55
Answer A is wrong. If the ground wire was loose or broken, then the blower motor would not work at all.
**Answer B is correct.** Only Technician B is correct. The problem is a blower motor that doesn't work in the high-speed mode. The cause must be something in that circuit and something that would affect the operation of the motor in only that switch position. If the high-speed relay were defective, only high-speed operation would be affected. The blower motor often receives its power through a high blower relay or a low blower relay. To determine when a blower motor is inoperative, it is necessary to bypass the other components in the circuit.
Answer C is wrong. Only Technician B is correct.
Answer D is wrong Only Technician B is correct.

Question #56

Answer A is wrong. A plugged evaporator would cause a greater temperature difference than the one described by this question.

**Answer B is correct.** Using the sense of touch is a good way to diagnose problems within the A/C and heating systems. The temperature of an A/C system's hoses indicates the condition of the system. By combining the results of both the hands-on checks and an interpretation of pressure gauge readings, the technician has a good indication that some unit in the system is malfunctioning and that further diagnosis is needed.. Equal temperatures at the inlet and outlet of the evaporator indicate the system is working normally and to do that, it must have the correct charge of refrigerant in the system.

Answer C is wrong. A restricted orifice tube would cause the outlet tube to be warmer than the inlet tube.

Answer D is wrong. A leaking condenser would cause system pressures to be lower than normal and the incorrect temperatures to be felt throughout the system.

Question #57

Answer A is wrong. The front-seated position traps refrigerant in the lines and allows for the compressor to be removed without the loss of refrigerant.

Answer B is wrong. The back-seated position is where the valve should be adjusted when testing and service is complete.

Answer C is wrong. The mid-seated position is where the valve should be when pressure testing the system. This is also the position for recovery or charging the system.

**Answer D is correct.** The A position is for isolating the compressor and sealing refrigerant in the system. Once refrigerant is recovered from the compressor it can be replaced or repaired, or the oil level can be checked. Position B is the normal operation position. A gauge connected at port 1 would not show any pressure. Position C is the testing and servicing position for the system. A gauge connected at port 1 would read system pressure. D is the EXCEPT answer. The system can be operated with a gauge connected to the 1 port to read system pressures. The valve should be returned to the back seated (B) position before removing the gauge.

Question #58

Answer A is wrong. Technician B is also correct.

Answer B is wrong. Technician A is also correct.

**Answer C is correct.** Both technicians are correct. Since R-12 is sold in white containers and R-134A is marketed in light blue containers.

Answer D is wrong. Both technicians are correct.

Question #59

**Answer A is correct.** A low refrigerant charge may cause faster-than-normal clutch cycling without frosting of any components. That is exactly what the problem is, so low charge is a very likely cause of the problem.

Answer B is wrong. A flooded evaporator would cause elevated low-side pressures and frost to develop on the suction line.

Answer C is wrong. A restricted accumulator would likely result in a noticeable temperature drop from the inlet to the outlet with the outlet developing frost.

Answer D is wrong. An overcharge of refrigerant would cause system pressures to be higher than normal.

Question #60

Answer A is wrong. Technician B is also correct.

Answer B is wrong. Technician A is also correct.

**Answer C is correct.** Both technicians are correct. The problem is that the A/C system seems to work fine in the early morning and evenings, but does not during the heat of the day. During a performance test, the low-side gauge reads normal and then drops dramatically. Both technicians are right. Moisture in the system can be released from the desiccant in the receiver/drier by an increase in temperature on the high side. When the ambient temperature is lower the desiccant may be able to hold the moisture out of the refrigerant. Internal freezing of the expansion valve could produce external icing on the valve.

Answer D is wrong. Both technicians are correct.

Question #61

Answer A is wrong. It would be dangerous to charge the system into the high side with the compressor running.

**Answer B is correct.** To charge a system with the engine running, the refrigerant should be added to the low side of the system. Never open the high-side hand valve with the system operating and a refrigerant can connected to the center hose. The refrigerant will flow out of the system under high pressure into the can. High-side pressure is between 150 and 300 psi and will cause the refrigerant tank to burst. The only occasion for opening both hand valves at the same time would be when evacuating the system or when reclaiming refrigerant with the system off.

Answer C is wrong. It would be dangerous to charge the system into the high side with the compressor running.

Answer D is wrong. It would be dangerous to charge the system into the high side with the compressor running.

Question #62

Answer A is wrong. A restricted cabin air filter would cause less heat to get absorbed into the refrigerant thus lowering system pressures.

**Answer B is correct.** The purpose of the cabin air filter is to remove pollen, dust, and other contaminates from the air entering the air distribution system. As with any air filter that is not serviced regularly, it will become restricted as it removes debris from the airflow. Answer B is correct; this will result in decreased airflow from the outlets.

Answer C is wrong. A restricted cabin air filter would not cause a frozen evaporator core. The de-icing devices should keep the evaporator from freezing even if the cabin air filter gets restricted.

Answer D is wrong. A stopped up evaporator drain would be caused by debris and dirt getting into the HVAC ducts. The cabin air filter would prevent any debris from getting to the evaporator core.

Question #63

Answer A is wrong. A refrigerant identifier should always be used before hooking any equipment up a vehicle. This would prevent accidentally getting mixed refrigerant into the shop's gauge sets and recovery machines.

Answer B is wrong. A refrigerant identifier should always be used before hooking any equipment up a vehicle. This would prevent accidentally getting mixed refrigerant into the shop's gauge sets and recovery machines.

Answer C is wrong. Neither technician is correct.

**Answer D is correct.** Neither technician is correct. In this question is correct. Due to the possibility of cross contaminants of refrigerants, the first thing that an A/C technician should do is analyze the refrigerant with the use of a refrigerant identifier. If technician A connects a manifold and gauge set to a system that has contaminated refrigerant, the refrigerant in the manifold and hoses could contaminate the next system the gauges are used on. If he recovers the refrigerant from the manifold and gauge set, as he should, he could contaminate the refrigerant in the recovery tank. If Technician B recovers contaminated refrigerant, all of the refrigerant in the recovery tank becomes useless. This could be a costly error.

Question #64

Answer A is wrong. Technician B is also correct.

Answer B is wrong. Technician A is also correct.

**Answer C is correct.** Both technicians are correct. Thermal switches are used on both the low and high side of AC systems. Technician A is right; a thermal switch may be connected in series with the compressor clutch. A thermostatic switch de-energizes the clutch and stops the compressor to keep the evaporator from freezing up. When the temperature of the evaporator approaches the freezing point (or the low setting of the switch), the thermostatic switch opens the circuit and disengages the compressor clutch. The compressor remains inoperative until the evaporator temperature rises to the preset temperature, at which time the switch closes and compressor operation resumes. Technician B is also right. A thermal limit switch may be mounted on or near the compressor and is used to stop compressor operation before the temperature of the compressor reaches a point of damage.

Answer D is wrong. Both technicians are correct.

Question #65
Answer A is wrong. Technician B is also correct.
Answer B is wrong. Technician A is also correct.
**Answer C is correct.** Both technicians are correct. There are many variations of electric cooling fans used on today's vehicles. Their operation depends on a number of things and they are controlled by a number of switches, relays, and control modules. Some cooling fan systems use a two-fan system; they have a low-speed fan and a high-speed fan. An electric cooling fan usually is wired through a relay because of the amount of current drawn to run an electric fan. Using a relay reduces the amount of current at the controlling switches. This reduces the chance of switch failure caused by arcing.
Answer D is wrong. Both technicians are correct.

Question #66
Answer A is wrong. Technician B is also correct.
Answer B wrong. Technician A is also correct.
**Answer C is correct.** Both technicians are correct. Some actuator motors are calibrated automatically in the self-diagnostic mode of the ECM, whereas other actuator motors must be calibrated manually. These actuators should only require calibration after motor replacement or misadjustment.
Answer D is wrong. Both technicians are correct.

Question #67
Answer A is wrong. Technician B is also correct.
Answer B is wrong. Technician B is also correct.
**Answer C is correct.** Both technicians are correct. The evaporator water drain is an integral part of the evaporator housing assembly. A crack or break in the housing may reduce the cooling capacity of the air conditioner. If the evaporator (water) drain becomes clogged, it is easily cleaned with a stiff wire from the outlet in, provided the technician takes care not to puncture the delicate evaporator core. If the case has a crack or break on the engine side of the firewall, excessive engine heat may enter the case and thereby reduce the efficiency of the air conditioning system.
Answer D is wrong. Both technicians are correct.

Question #68
Answer A is wrong. Using an ammeter to look for a short is not recommended because the internal fuse on the meter would likely blow due to high current flow.
Answer B is wrong. The only load in the A/C compressor circuit is the coil, so performing voltage drop tests would not help find the problem.
**Answer C is correct.** The best way to find a short in a compressor clutch coil is by using an ohmmeter to measure the resistance of the coil. The normal resistance is usually only a few ohms so it is important to compare your reading to the specification for your compressor.
Answer D is wrong. The specifications for the magnetic strength of a clutch coil are not published in service literature.

Question #69
Answer A is wrong. Air in the system would cause elevated system pressures.
Answer B is wrong. A bad compressor would cause high low-side pressures and low high-side pressures.
**Answer C is correct.** A low refrigerant charge causes reduced low-side and high-side pressures. A restriction in the low side will cause lower-than-normal low-side pressure, even into a vacuum. Such a restriction in the low side of the system would generally be accompanied by a lower-than-normal high-side pressure. This is also true if there is an undercharge of refrigerant. If the system is overcharged, however, the higher-than-normal high-side pressure will be accompanied with a higher-than-normal low-side pressure.
Answer D is wrong. A stuck-open TXV would cause elevated pressures on the low-side of the system and frosting on the suction line.

Question #70

**Answer A is correct.** Only Technician A is correct. Part of the retrofit process is to install a new label that includes what oil was used in the retrofit.
Answer B is wrong. Several different viscosities and blends of PAG oils are used by different manufacturers. The technician should always consult the service manual and use the recommended oil.
Answer C is wrong. Only Technician A is correct.
Answer D is wrong. Only Technician A is correct.

Question #71

Answer A is wrong. An excessive oil charge would cause elevated system pressures and warmer duct temperatures but it would not cause oil to leak out of hose and line connections.
Answer B is correct. Only Technician B is correct. An oily residue indicates a refrigerant leak in that area where refrigerant is escaping along with some of the oil in the system.
Answer C is wrong. Only Technician B is correct.
Answer D is wrong. Only Technician B is correct.

Question #72

Answer A is correct. Only Technician A is correct. These gauge readings are too high. The cause of the readings could be air and moisture in the refrigerant system or an overcharged system. A restricted thermal expansion valve (TXV) may cause low refrigerant pressures; this condition may also result in TXV frosting.
Answer B is wrong. A restricted TXV would cause low system pressures.
Answer C is wrong. Only Technician A is correct.
Answer D is wrong. Only Technician A is correct.

Question #73

Answer A is wrong. A metering device that is stuck closed would cause low pressure on the low side of the A/C system.
Answer B is wrong. Poor airflow across the evaporator would reduce the heat load and therefore reduce the low side operating pressure.
Answer C is wrong. A restriction in the low side of the system would cause the pressure to drop in the low side of the A/C system.
**Answer D is correct.** If the low-side pressure is lower than normal, suspect a faulty metering device, poor airflow across the evaporator, a restriction in the low side of the system, or the system is *under*charged with refrigerant. D is the Least-Likely answer because an overcharge results in both pressures being high.

# Glossary

**Access valve** A term used for service port and service valve.

**Accumulator** A tank located between the evaporator and compressor to receive the refrigerant that leaves the evaporator, so constructed as to ensure that no liquid refrigerant will enter the compressor.

**Actuator** A device that transfers a vacuum or electric signal to a mechanical motion, typically performs an on/off or open/close function.

**Adapter** A device or fitting that permits different size parts or components to be fastened or connected to each other.

**Aftermarket** A term given to a device or accessory that is added to a vehicle after original manufacture, such as an air conditioning system.

**Air gap** The space between two components, such between the rotor and armature of a clutch.

**Ambient sensor** A thermistor used in automatic temperature control units to sense ambient temperature.

**Armature** The part of the clutch that mounts onto the crankshaft and engages with the rotor when energized.

**Atmospheric pressure** Air pressure at a given altitude. At sea level, atmospheric pressure is 14.696 psia (101.329 kPa absolute).

**Back seat (service valve)** Turning the valve stem to the left (ccw) as far as possible back seats the valve. The valve outlet to the system is open and the service port is closed.

**Barb fitting** A fitting that slips inside a hose and is held in place with a gear-type clamp. Ridges (barbs) on the fitting prevent the hose from slipping off.

**BCM** An abbreviation for body control module.

**Blower relay** An electrical device used to control the function or speed of a blower motor.

**Boiling point** The temperature at which a liquid changes to a vapor.

**Break a vacuum** The next step after evacuating a system. The vacuum should be broken with refrigerant or other suitable dry gas, not ambient air or oxygen.

**Bypass** An alternate passage that may be used instead of the main passage.

**By-pass hose** A hose that is generally small and is used as an alternate passage to bypass a component or device.

**CAA** Clean Air Act.

**Can tap** A device used to pierce, dispense, and seal small cans of refrigerant.

**Can tap valve** A valve found on a can tap that is used to control the flow of refrigerant.

**Cap** (1) A protective cover. (2) An abbreviation for capillary (tube) or capacitor.

**Cap tube** A tube with a calibrated inside diameter and length used to control the flow of refrigerant such as rgat between the remote bulb to the expansion valve.

**Celsius (C)** A metric temperature scale using zero as the freezing point of water. The boiling point of water is 100° C (212° F).

**Certified** Having a certificate awarded or issued to those that have demonstrated appropriate competence through testing and/or practical experience.

**CFC-12** A term used for Refrigerant-12.

**Charge** A specific amount of refrigerant or oil by volume or weight.

**Check valve** A device that prevents refrigerant from flowing in the opposite direction when the unit is shut off.

**Clearn Air Act (CAA)** A Title IV amendment signed into law in 1990 which established national policy relative to the reduction and elimination of ozone-depleting substances.

**Clockwise (cw)** A left to right rotation or motion.

**Clutch** An electro-mechanical device mounted on the air conditioning compressor used to start and stop compressor action, thereby controlling refrigerant circulating through the system.

**Clutch coil** The electrical part of a clutch assembly. When electrical power is applied to the clutch coil, the clutch is engaged to start and stop compressor action.

**Compound gauge** A gauge that registers both pressure and vacuum (above and below atmospheric pressure); used on the low side of the systems.

**Compressor-shaft seal** An assembly consisting of springs, snap rings, O-rings, shaft seal, seal sets, and gasket, mounted on the compressor crankshaft to permit the shaft to be turned without a loss of refrigerant or oil.

**Contaminated** A term used when referring to a refrigerant cylinder or a system that is known to contain foreign substances such as other incompatible or hazardous refrigerants.

**Counterclockwise (ccw)** A direction, right to left, opposite that which a clock turns.

**Cracked position** A mid-seated or open position.

**Cycle clutch time (total)** Time elapsed from the moment the clutch engages until it disengages, then reengages. Total time is equal to on-time plus off-time for one cycle.

**Cycling clutch pressure switch** A pressure-actuated electrical switch used to cycle the compressor at a predetermined pressure.

**Cycling clutch system** An air conditioning system in which the air temperature is controlled by starting and stopping the compressor with a thermostat or pressure control.

**Department of Transportation (DOT)** A federal agency charged with regulation and control of the shipment of all hazardous materials.

**Depressing pin** A pin located in the end of a service hose to press (open) a Schrader-type valve.

**Disarm** To turn off; to disable a device or circuit.

**Dry nitrogen** The element nitrogen (N) which has been processed to ensure that it is free of moisture.

**Dual system** Two systems; usually refers to two evaporators in an air conditioning system; one in the front and one in the rear of the vehicle, driven off a single compressor and condenser system.

**Duct** A tube or passage used to provide a means to transfer air or liquid from one point or place to another.

**EATC** Electronic Automatic Temperature Control.

**ECC** Electronic Climate Control.

**Enviornmental Protection Agency (EPA)** An agency of the U.S. government that is charged with the responsibility of protecting the environment and enforcing the Clean Air Act (CAA) of 1990.

**EPA** Environmental Protection Agency.

**Etch** An intentional or unintentional erosion of a metal surface generally caused by an acid.

**Evacuate** To create a vacuum within a system to remove all traces of air and moisture.

**Evaporator core** The tube and fin assembly located inside the evaporator housing. The refrigerant fluid picks up heat in the evaporator core when it changes into a vapor.

**Expansion tank** An auxiliary tank that is usually connected to the inlet tank of a radiator and which provides additional storage space for heated coolant. Often called a coolant recovery tank.

**External snap ring** A snap ring found on the outside of a part such as a shaft.

**Fan relay** A relay for the cooling and/or auxiliary fan motors.

**Fill neck** The part of the radiator on which the pressure cap is attached. Most radiators, however, are filled via the recovery tank.

**Filter** A device used with the dryer or as a separate unit to remove foreign material from the refrigerant.

**Filter dryer** A device that has a filter to remove foreign material from the refrigerant and a desiccant to remove moisture from the refrigerant.

**Flare** A flange or cone-shaped end applied to a piece of tubing to provide a means of fastening to a fitting.

**Forced air** Air that is moved mechanically such as by a fan or blower.

**Front seat** Closing off the line leaving the compressor open to the service port fitting. This allows service to the compressor without purging the entire system. Never operate the system with the valves front seated.

**Functional test** A term used for performance test.

**Fusible link** A type of fuse made of a special wire that melts to open a circuit when current draw is excessive.

**Gasket** A thin layer of material or composition that is placed between two machined surfaces to provide a leakproof seal between them.

**Gauge** A tool of a known calibration used to measure components. For example, a feeler gauge is used to measure the air gap between a clutch rotor and armature.

**Graduated container** A measure such as a beaker or measuring cup that has a graduated scale for the measure of a liquid.

**Ground** A general term given to the negative (–) side of an electrical system.

**Grounded** An intentional or unintentional connection of a wire, positive (+) or negative (–), to the ground. A short circuit is said to be grounded.

**Gross weight** The weight of a substance or matter that includes the weight of its container.

**HCFC** Hydrochlorofluorocarbon refrigerant.

**Header tank** The top and bottom tanks (downflow) or side tanks (crossflow) of a radiator. The tanks in which coolant is accumulated or received.

**Heater core** A radiator-like heat exchanger located in the case/duct system through which coolant flows to provide heat to the vehicle interior.

**Heat exchanger** An apparatus in which heat is transferred from one medium to another on the principle that heat moves to an object with less heat.

**HI** The designation for high as in blower speed or system mode.

**High-side gauge** The correct side gauge on the manifold used to read refrigerant pressure in the high side of the system.

**High-side hand valve** The high-side valve on the manifold set used to control flow between the high side and service ports.

**High-side service valve** A device located on the discharge side of the compressor; this valve permits the service technician to check the high-side pressures and perform other necessary operations.

**High-side switch** See Pressure switch.

**High-torque clutch** A heavy-duty clutch assembly used on some vehicles known to operate with higher-than-average head pressure.

**Hot knife** A knife-like tool that has a heated blade used for separating objects, such as evaporator cases.

**Hub** The central part of a wheel-like device such as a clutch armature.

**Hygiene** A system of rules and principles intended to promote and preserve health.

**Hygroscopic** Readily absorbing and retaining moisture.

**Idler** A pulley device that keeps the belt whip out of the drive belt of an automotive air conditioner. The idler is used as a means of tightening the belt.

**Idler pulley** A pulley used to tension or torque the belt(s).

**Idle speed** The speed (rpm) at which the engine runs while at rest (idle).

**In-car temperature sensor** A thermistor used in automatic temperature control units for sensing the in-car temperature. Also see Thermistor.

**Insert fitting** A fitting that is designed to fit inside, such as a barb fitting that fits inside a hose.

**Internal snap ring** A snap ring used to hold a component or part inside a cavity or case.

**Jumper** A wire used to temporarily bypass a device or component for the purpose of testing.

**Kilogram (kg)** A unit of measure in the metric system. One kilogram is equal to 2.205 pounds in the English system.

**KiloPascal (kPa)** A unit of measure in the metric system. One kilopascal (kPa) is equal to 0.145 pound per square inch (psi) in the English system.

**kPa** kiloPascal.

**Liquid** A state of matter; a column of fluid without solids or gas pockets.

**Low-refrigerant switch** A switch that senses low pressure due to a loss of refrigerant and stops compressor action. Some alert the operator and/or set a trouble code.

**Low-side gauge** The left-side gauge on the manifold used to read refrigerant pressure in the low side of the system.

**Low-side hand valve** The manifold valve used to control flow between the low side and service ports of the manifold.

**Low-side service valve** A device located on the suction side of the compressor which allows the service technician to check low-side pressures and perform other necessary service operations.

**Manifold** A device equipped with a hand shutoff valve. Gauges are connected to the manifold for use in system testing and servicing.

**Manifold and gauge set** A manifold complete with gauges and charging hoses.

**Manifold hand valve** Valves used to open and close passages through the manifold set.

**Manufacturer's procedures** Specific step-by-step instructions provided by the manufacturer for the assembly, disassembly, installation, replacement, and/or repair of a particular product manufactured by them.

**MAX** A mode, maximum, for heating or cooling. Selecting MAX generally overrides all other conditions that may have been programmed.

**Mid-positioned** The position of a stem-type service valve where all fluid passages are interconnected. Also referred to as "cracked."

**Motor** An electrical device that produces a continuous turning motion. A motor is used to propel a fan blade or a blower wheel.

**MSDS** Material Safety Data Sheet.

**Net weight** The weight of a product only; container and packaging not included.

**Neutral** On neither side; the position of gears when force is not being transmitted.

**Noncycling clutch** An electro-mechanical compressor clutch that does not cycle on and off as a means of temperature control; it is used to turn the system on when cooling is desired and off when cooling is not desired.

**OEM** Original equipment manufacturer.

**Off-the-road** Generally refers to vehicles that are not licensed for road use, such as harvesters, bulldozers, and so on.

**Ohmmeter** An electrical instrument used to measure the resistance in ohms of a circuit or component.

**Open** Not closed. An open switch, for example, breaks an electrical circuit.

**Orifice** A calibrated opening in a tube or pipe to regulate the flow of a fluid or liquid.

**O-ring** A synthetic rubber or plastic gasket with a round- or square-shaped cross-section.

**OSHA** Occupational Safety and Health Administration.

**Outside temperature sensor** A term used for ambient sensor.

**Overcharge** Indicates that too much refrigerant or refrigeration oil is added to the system.

**Overload** Anything in excess of the design criteria. An overload will generally cause the protective device such as a fuse or pressure relief to open.

**Ozone friendly** Any product that does not pose a hazard or danger to the ozone.

**Park** Generally refers to a component or mechanism that is at rest.

**PCM** Power train control module.

**Performance test** Readings of the temperature and pressure under controlled conditions to determine if an air conditioning system is operating at full efficiency.

**Piercing pin** The part of a saddle valve that is used to pierce a hole in the tubing.

**Pin-type connector** A single or multiple electrical connector that is round- or pin-shaped and fits inside a matching connector.

**Poly belt** A term used for serpentine belt.

**Polyester (ESTER)** A synthetic oil-like lubricant that is occasionally recommended for use in an HFC-134a system. This lubricant is compatible with both HFC-134a and CFC-12.

**Positive pressure** Any pressure above atmospheric.

**Pound** A weight measure, 16 ounces.

**Pound of refrigerant** A term often used by technicians when referring to a small can of refrigerant although it actually contains less than 16 ounces.

**Power module** Controls the operation of the blower motor in an automatic temperature control system.

**Predetermined** A set of fixed values or parameters that have been programmed or otherwise fixed into an operating system.

**Pressure gauge** A calibrated instrument for measuring pressure.

**Pressure switch** An electrical switch that is activated by a predetermined low or high pressure. A high-pressure switch is generally used for system protection; a low pressure switch may be used for temperature control or system protection.

**Propane** A flammable gas used as a propellant for the halide leak detector.

**PSIG** Pounds per square inch gauge.

**Purge** To remove moisture and/or air from a system or a component by flushing with a dry gas such as nitrogen (N) to remove all refrigerant from the system.

**Purity test** A static test that may be performed to compare the suspect refrigerant pressure to an appropriate temperature chart to determine its purity.

**Radiation** The transfer of heat without heating the medium through which it is transmitted.

**Ram air** Air that is forced through the radiator and condenser coils by the movement of the vehicle or the action of the fan.

**Receiver/drier** A tank-like vessel having a desiccant and used for the storage of refrigerant.

**RECIR** An abbreviation for the recirculate mode, as with air.

**Recovery system** A term often used to refer to the circuit inside the recovery unit used to recycle and/or transfer refrigerant from the air conditioning system to the recovery cylinder.

**Recovery tank** An auxiliary tank, usually connected to the inlet tank of a radiator, which provides additional storage space for heated coolant.

**Refrigerant-12** The refrigerant used in automotive air conditioners, as well as other air conditioning and refrigeration systems.

**Relay** An electrical switch device that is activated by a low-current source and controls a high-current device.

**Reserve tank** A storage vessel for excess fluid. See Recovery tank, Receiver/drier, and Accumulator.

**Resistor** A voltage-dropping device that is usually wire wound and provides a means of controlling fan speeds.

**Respirator** A mask or face shield worn in a hazardous environment to provide clean fresh air and/or oxygen.

**Restrictor** An insert fitting or device used to control the flow of refrigerant or refrigeration oil.

**Retrofitting** The name given to the procedure for converting R-12 A/C systems to be able to use R-134a refrigerant.

**Rotor** The rotating or freewheeling portion of a clutch; the belt slides on the rotor.

**Rpm** Revolutions per minute; also, rpm or r/min.

**Running design change** A design change made during a current model/year production.

**Saddle valve** A two-part accessory valve that may be clamped around the metal part of a system hose to provide access to the air conditioning system for service.

**SAE** Society of Automotive Engineers.

**Schrader valve** A spring-loaded valve similar to a tire valve. The Schrader valve is located inside the service valve fitting and is used on some control devices to hold refrigerant in the system. Special adapters must be used with the gauge hose to allow access to the system.

**Seal** Generally refers to a compressor shaft oil seal; matching shaft-mounted seal face and front head-mounted seal seat to prevent refrigerant and/or oil from escaping. May also refer to any gasket or O-ring used between two mating surfaces for the same purpose.

**Seal seat** The part of a compressor shaft seal assembly that is stationary and matches the rotating part, known as the seal face or shaft seal.

**Serpentine belt** A flat or V-groove belt that winds through all of the engine accessories to drive them off the crankshaft pulley.

**Service port** A fitting found on the service valves and some control devices; the manifold set hoses are connected to this fitting.

**Service procedure** A suggested routine for the step-by-step act of troubleshooting, diagnosing and/or repairs.

**Service valve** See High-side (Low-side) service valve.

**Shaft key** A soft metal key that secures a member on a shaft to prevent it from slipping.

**Shaft seal** See Compressor-shaft seal.

**Short of brief duration** e.g., short cycling. Also refers to an intentional or unintentional grounding of an electrical circuit.

**Shut-off valve** A valve that provides positive shut-off of a fluid or vapor passage.

**Snap ring** A metal ring used to secure and retain a component to another component.

**Society of Automotive Engineers** A professional organization of the automotive industry. Founded in 1905 as the Society of Automobile Engineers, the SAE is dedicated to providing technical information and standards to the automotive industry.

**Solenoid valve** An electromagnetic valve controlled remotely by electrically energizing and deenergizing a coil.

**Solid state** Referring to electronics, consisting of semiconductor devices and other related nonmechanical components.

**Spade-type connector** A single or multiple electrical connector that has flat spade-like mating provisions.

**Specifications** Design characteristics of a component or assembly noted by the manufacturer. Specifications for a vehicle include fluid capacities, weights, and other pertinent maintenance information.

**Spike** In our application, an electrical spike. An unwanted momentary high-energy electrical surge.

**Spring lock fitting** A special fitting using a spring to lock the mating parts together forming a leak-proof joint.

**Squirrel-cage blower** A blower wheel designed to provide a large volume of air with a minimum of noise. The blower is more compact than the fan and air can be directed more efficiently.

**Stabilize** To make steady.

**Stratify** Arrange or form into layers. To fully blend.

**Subsystem** A system within a system.

**Sun load** Heat intensity and/or light intensity produced by the sun.

**Superheat switch** An electrical switch activated by an abnormal temperature-pressure condition (a superheated vapor); used for system protection.

**Tank** A term used for header tank and expansion tank.

**Tare weight** The weight of the packaging material.

**Temperature door** A door within the case/duct stem to direct air through the heater and/or evaporator core.

**Temperature switch** A switch actuated by a change in temperature at a predetermined point.

**Tension gauge** A tool for measuring the tension of a belt.

**Thermistor** A temperature-sensing resistor that has the ability to change values with changing temperature.

**Torque** A turning force; for example, the force required to seal a connection; measured in (English) foot-pounds (ft-lb) or inch-pounds (in-lb); (metric) Newton-meters (N·m).

**Triple evacuation** A process of evacuation that involves three pump-downs and two system purges with an inert gas such as dry nitrogen (N).

**TXV** Thermostatic expansion valve.

**Ultraviolet (uv)** The part of the electromagnetic spectrum emitted by the sun that lies between visible violet light and X-rays.

**Vacuum gauge** A gauge used to measure below atmospheric pressure.

**Vacuum motor** A device designed to provide mechanical control by the use of a vacuum.

**Vacuum pump**  A mechanical device used to evacuate the refrigeration system to rid it of excess moisture and air.

**Vacuum signal**  The presence of a vacuum.

**V-belt**  A rubber-like continuous loop placed between the engine crankshaft pulley and accessories to transfer rotary motion of the crankshaft to the accessories.

**Ventilation**  The act of supplying fresh air to an enclosed space such as the inside of an automobile.

**V-groove belt**  A term used for V-belt.

**Voltmeter**  A device used to measure volt(s).

**Wiring harness**  A group of wires wrapped in a shroud for the distribution of power from one point to another point.

# Notes

Notes

**Notes**

CPSIA information can be obtained
at www.ICGtesting.com
Printed in the USA
FFOW020800300113
783FF

9 781418 038847